The Environmental Impact of Land and Property Management

The Environmental Impact of Land and Property Management

Edited by

YVONNE RYDIN
London School of Economics, UK

Published on behalf of
The Royal Institution of Chartered Surveyors (RICS)

by

JOHN WILEY & SONS
Chichester · New York · Brisbane · Toronto · Singapore

Copyright © 1996 The Royal Institution of Chartered Surveyors

Published 1996 by John Wiley & Sons Ltd,
Baffins Lane, Chichester,
West Sussex PO19 1UD, England

National   01243 779777
International   (+44) 1243 779777
e-mail (for orders and customer service enquiries): cs-books@wiley.co.uk
Visit our Home Page on http://www.wiley.co.uk
or http://www.wiley.com

Other Wiley Editorial Offices

John Wiley & Sons, Inc., 605 Third Avenue,
New York, NY 10158-0012, USA

Jacaranda Wiley Ltd, 33 Park Road, Milton,
Queensland 4064, Australia

John Wiley & Sons (Canada) Ltd, 22 Worcester Road,
Rexdale, Ontario M9W 1L1, Canada

John Wiley & Sons (Asia) Pte Ltd, 2 Clementi Loop #02-01,
Jin Xing Distripark, Singapore 0512

Library of Congress Cataloging-in-Publication Data

Rydin, Yvonne, 1957–
      The environmental impact of land and property management / Yvonne
Rydin.
            p.   cm.
      Includes bibliographical references and index.
      ISBN 0-471-96612-6
      1. Land use—Great Britain—Planning.   2. Real estate management—
Great Britain—Planning.   3. Land use—Environmental aspects—Great
Britain.   4. Real estate management—Environmental aspects—Great
Britain.
      HD596.R93   1996
      333.73'0941—dc20   96–19504
                                                                                    CIP

British Library Cataloguing in Publication Data

A catalogue record for this book is available from the British Library

ISBN 0-471-96612-6

Set from CRC supplied by RICS
Printed and bound in Great Britain by Biddles Ltd, Guildford and King's Lynn
This book is printed on acid-free paper responsibly manufactured from sustainable forestation, for which at least two trees are planted for each one used for paper production.

Contents

List of Contributors

Dr Yvonne Rydin BA(Hons), ARICS, Department of Geography, London School of Economics

Sue Markwell BSc(Hons) and **Dr Neil Ravenscroft** BSc(Hons), MSc, ARICS, DipRM, Centre for Environmental and Land Tenure Studies, University of Reading

Alison Darlow BSc(Hons) and **Norma Carter** BA(Hons), DipTP, MRTPI, Department of Land Management, De Montfort University

Dr Christine Pasquire BSc(Hons), FRICS, Department of Civil Engineering and Surveying, Loughborough University of Technology

Mark Bhatti and **Chris Sarno**, Centre for Local Environmental Policy and Strategies, South Bank University

Sandra Jane Dixon MSc, ARICS, School of the Built Environment, University of Northumbria at Newcastle

Dr Timothy J. Dixon BA(Hons), DipDistEd, FRICS and **Tim Richards** BSc(Hons), MSc, College of Estate Management

Jane Taussik BA(Hons) DipTP, MRTPI, Department of Land and Construction Management, University of Portsmouth

Dr A.R. Ghanbari Parsa, School of Land Management and Urban Policy, South Bank University

Dr Geoffrey Barker-Read BSc(Hons), MSc MIMinE, CEng, FMVSSA and **Robert A. Farnfield** BSc, MIExpE, Department of Mining and Mineral Engineering, University of Leeds

Preface

In 1994 The Royal Institution of Chartered Surveyors commissioned a programme of nine research projects which aimed to address a wide spectrum of environmental issues being faced by the surveying profession. The work was undertaken in the context of increasing awareness of "green issues" and of environmental responsibility amongst the population as a whole. The researchers involved presented their work at the Environmental Research Conference held at The Royal Institution of Chartered Surveyors headquarters in November 1995. This book is a compilation of the findings of the research presented at the conference.

The unifying theme running through the book is a recognition of the need for improved environmental education amongst the property profession. As the leading suppliers of professional services to the land, property and construction industries, chartered surveyors have a greater influence on environmental decision-making than many other groups. Therefore, it is imperative that property professionals aim to develop the appropriate skills and knowledge of key environmental issues. The Royal Institution of Chartered Surveyors accepts that it has a major role to play in achieving this aim, and is committed to promoting environmental improvement and sustainable development.

Acknowledgements

The RICS wishes to thank the following people for their help with the Environmental Research Programme:

Environmental Research Programme Co-ordinator
Dr Yvonne Rydin, London School of Economics

the members of the steering group for the project including
Charles Cowap FRICS, Harper Adams Agricultural College
Hugh Cross FRICS, Matthews & Goodman, London
Karen Sieracki ARICS, ESN Pension Management Group
Philip Wilbourn FRICS, Philip Wilbourn Associates, Sheffield
Alison Tanner, RICS Secretariat

1 Introduction: The Environmental Impact of Land and Property Management

YVONNE RYDIN
London School of Economics, London, UK

THE NEW ENVIRONMENTAL AGENDA

The later 1980s and early 1990s have seen a wave of environmental concern quite distinct from that of the 1960s and early 1970s. In the earlier period, environmentalism was seen as a fringe concern, associated with radical social movements and marginal lifestyles (Schumacher, 1973). Now everyone makes claims to be 'green': corporations advertise their concern for the environment, government policy repeatedly emphasises stewardship and conservation and forums for green business, green education, green local authorities thrive. How did this shift in public and private consciousness come about?

The event that brought the first wave of environmentalism to a close was undoubtedly the world economic recession following the 1973/4 oil price rises and the Arab-Israeli war. Concern with economic growth, employment and real incomes made concern with the environment seem a luxury. Ironically the energy price increase not only brought the threat of energy and resource shortages home to many, it also led to an increase in energy efficiency. And the reduced demand consequent upon a slow down in economic growth took the threat of resource shortages further away. So that environmentalism not only seemed a luxury, it was an unnecessary luxury. Environmentalism moved even further away from the mainstream political agenda.

Then in the 1980s scientific evidence on first the growth of the ozone hole over the polar regions, and then the rise in global temperatures known as global warming, both began to have a widespread effect on how we viewed the environment (IPCC, 1990). Risks associated with environmental change again seemed real and urgent. Furthermore we began to see the relationship between economic growth and environmental protection in different terms. This was no longer a simple relation of trade-offs in which we could either have development

The Environmental Impact of Land and Property Management, Edited by Yvonne Rydin
©1996, The Royal Institution of Chartered Surveyors

or environmental concern. Economic activities were seen as dependent on the environmental functions that we tend to take for granted and maintaining living standards would involve rethinking our use of those functions (Pearce, 1989).

The policy catalyst was undoubtedly the report of the United Nations Commission on Environment and Development, the Brundtland Report (WCED, 1987). This popularised the notion of 'sustainable development', the idea that we should not meet our own needs by compromising the ability of future generations to meet their needs. Central to this idea was the recognition that we are currently using up environmental capital, in terms of both stock natural resources and also the earth's ability to recycle our production of waste products. We are consuming minerals and fossil fuels; we are destroying habitats and the genetic stock of biodiversity; we are polluting land, water and air media; and we are unbalancing environmental systems, both locally as with deforestation destroying watershed protection and globally as with disruption of the stratosphere and climate control functions. And these changes may be irreversible, fundamentally threatening human beings' ability to survive: whether endangering individual lives, threatening local communities or even, possibly, the prospects of the human species (Lovelock, 1983).

This new recognition of environmental threats has had a profound effect. Public opinion in developed countries appears to register a new level of green concern (Worcester, 1993). This is not a consistent level of concern, still being vulnerable to an overriding preoccupation with economic prosperity, and it is not based on a particularly strong understanding of the scientific processes underpinning environmental change (Burgess, 1990 and Kempton, 1991). Nevertheless in developed countries it now appears that environmental considerations are a factor that the governments need to take into account in dealing with their electorates. At the same time the ongoing Brundtland process continues to have an effect.

The follow-up conference in Rio in 1992, the UN Conference on Environment and Development more popularly known as 'the Earth Summit', led to some modest successes in agreeing international environmental treaties (Grubb et al., 1993). It also led to the agreement by governments to prepare strategies for achieving this goal and monitoring government progress to this end. At the European level, the 5th environmental action programme is entitled 'Towards Sustainability' (1993) and at the British level, the government has issued 'Sustainable Development - the UK strategy' (1994). Local government has been particularly active, perhaps seeing in local environmental action a new role for beleaguered local authorities in a post-Thatcher era. The result has been a strong Local Agenda 21 process (see Darlow and Carter's paper in this volume), a wave of local sustainable development strategies and many individual initiatives.

So this book belongs to a growing process of environmental awareness affecting governments and the public, and the professions as well. It looks specifically at the way that the new environmental agenda relates to land and property management and the relevant professions. This relationship has an international dimension in two senses. First, as noted above, many environmental problems have a global dimension and have received policy attention at the global scale. Second, the interaction of the broader environment with land and property management is a relevant factor world-wide. However, environmental impacts and concerns vary considerably across the globe and, in particular, are quite distinct in the North and the South. Levels of poverty and economic development, severity of contamination and degradation, and the role of governments all differentiate the environmental agenda in different regions of the globe. Here, we deal with the environmental impact of land and property management in the developed world and in the policy context of Britain. This is not to deny the pressing need to achieve sustainable development in less developed countries. It simply reflects the need to start somewhere when tackling this problem, and one's own back yard seems as good a place as anywhere.

LAND AND PROPERTY MANAGEMENT AND THE ENVIRONMENT

It is an indicator of the rising importance of environmental issues that The Royal Institution of Chartered Surveyors (RICS) decided in 1994 to fund a research programme to consider the specific implications of the new environmental agenda for land and property management. This book brings together the results of the nine projects funded under that initiative. Together they provide a unique account of how environmental concerns already do and will continue in the future to play a part in land and property management.

The research reported here has emphasised the growing importance of these environmental concerns. There are three different kinds of issue at stake. First, there are the environmental consequences of the management of professional practices. This includes the generation of paper waste, the energy efficiency of their offices, the use of unleaded petrol in surveyors' cars and so on. As with other service industry businesses, adopting an in-house environmental strategy or even an environmental management and auditing system can focus attention on these consequences and ensure that the management of practices is not in itself a source of environmental problems. Markwell and Ravenscroft's chapter, for example, considers the changes needed for rural chartered surveying practices.

Secondly, there are the environmental consequences of the advice that land and property managers give to their clients. While the profession forms only a small proportion of the total service sector, the land and property sectors that such professionals advise are of considerable significance within the economy and the built or natural environment. The energy efficiency and waste management decisions of landowners and managers, their attitude to natural habitats and disruption of local ecosystems cumulatively amount to a significant impact. Changed professional advice could therefore cumulatively help to alter the relationship between environment and economy towards, and by implication also away from, the goal of sustainable development. For example, buildings account for about half of all energy consumption in Britain and the design and construction of these buildings is a major factor in determining the level of energy use. As the paper by Dixon explores, building surveying professionals are implicated in decisions about building construction and refurbishment which could cumulatively affect the overall energy consumption of the building stock.

Thirdly, and more generally, the land and property management profession is not just implicated in site specific management decisions. It performs a central role in the land and property markets as a whole. The actions of professionals constitute the exchange processes that together make a market. Professionals are active in regulating and enforcing market transactions. They also act at the interface of the market and the public sector, the economy and the state. As such they are central to the very concept of sustainable development. For this concept is not simply the expression of a pious policy hope; it embodies a powerful critique of how economic processes work and how governments relate to those processes.

Different analysts have taken up this critique in different ways: Pearce has developed the analysis of market and government failure to argue for an extended notion of total economic value incorporating the value of currently unpriced environmental goods and services (Pearce, 1989 and 1995); Daly has developed a more wide-ranging critique of Keynesian economics to argue that issues of the scale of production and consumption need to be reassessed to limit the throughput of matter (Daly, 1992) and Ekins has placed equity considerations at the heart of a reconceptualisation of economy-environment inter-relations by stressing the extent of non-market transactions particularly in less developed regions (Ekins and Max-Neef, 1992).

Therefore land and property management as a profession is implicated in these broader issues. The assumptions that managers adopt, the way in which they conduct exchanges, their treatment of non-market and unpriced aspects of land and property, all of these features underpin the type of sustainable development that can be achieved. For example Bhatti and Sarno's paper considers the role of estate agents and other exchange professionals in 'creating' green demand by managing information on energy efficiency in the new housing market.

All the research projects reported here provide a detailed insight into this three-fold significance of the land and property management profession to the environment. These are varied insights since the research projects covered the breadth of surveying practice, from planning and development, through land valuation, to mineral surveying and rural estate management. However, by reviewing this breadth it is possible to draw some overall conclusions on the prospects of change in the profession. The remainder of this introduction will address how such change may be achieved.

CHANGING VALUES

Sociological analyses of professions emphasise their nature as a form of occupational control (MacDonald, 1995). They are a way of organising working practices, of limiting access to a particular job of work and of socialising entrants into a common identity and mode of operation. By doing so, it is possible to charge a premium for the work and to position professionals more firmly within occupational and social structures. This is justified on the basis of the specialist expertise and command of knowledge that the professional has and the vocation of service held by the professional towards the client and the wider community. Therefore the knowledge, skills and values of the professional are central to establishing his or her role in society.

On this basis, change in professional practice can be achieved through altering these attributes of knowledge, skills and values. Indeed many of the papers in this volume emphasise the need for land and property managers to obtain certain basic information and knowledge on environmental impacts and trends, and on environmental policy and practice. Pasquire emphasises this in the case of quantity surveying, a sector of the profession which does not yet seem very environmentally aware. Skills in identifying the impacts arising from specific practices and learning how to mitigate them are further changes required. The educational process has a particularly important task to perform in providing new entrants into the profession with these skills, information and knowledge, as Parsa discusses.

But just acquiring these skills and knowledge is not enough. There has to be a positive impetus to using them. Focusing on the sociological organisation of the profession suggests that the value system of land and property management could be reoriented to encourage the individual professional to feel compelled to bring an environmental perspective to bear in his or her work, to raise potential environmental impacts and suggest changes in practice for the clients. Such a value shift is often cited as important in provoking changes in professional firms' own business practices. In encouraging this value shift, promoters of environmental concerns can point to the long term claim of the surveying profession to stewardship of land and property, a claim which itself implies a long term view and is now often used as a synonym for sustainable development. They can also cite the professional creed of service to the wider society as well as to the individual client.

These mechanisms can clearly work in the case of committed individuals. However a sceptical view would question the prospects for achieving widespread changes in individual professionals' values and the effectiveness of such value shifts in altering actual practice. Calls can be made for the professional body to enforce such shifts, as indeed is happening with the new guidelines from The Royal Institution of Chartered Surveyors, the 1995 Red Book, which makes consideration of environmental considerations mandatory. However relying solely or even mainly on the professional body in this way just shifts questions about commitment to environmental protection and impact on everyday practice to the institutional rather than the individual level. The question remains of how changed value systems in relation to the environment, whether individually motivated or institutionally enforced, can have an effect in the context of the prevailing economic considerations that professional firms face.

THE MARKET FOR LAND AND PROPERTY MANAGEMENT SKILLS

The reality for most professionals is that they need to sell their skills in the marketplace and will only be able to incorporate environmental concerns where it is viable to do so. Certain of the papers in this volume do argue just this, that surveyors as a profession can find a profitable market niche through offering environmental services: see, for example, Pasquire's paper. This may be due to the requirements of government policy and regulation imposing certain needs for environmental assessment and consideration; the question here becomes how surveyors compete with other professionals in trying to meet the skills gap created by such policy and regulation. Alternatively there may be economies to the client in obtaining specific knowledge about the environmental impact of their decisions and actions. With some recycling, waste management and energy efficiency measures there may be a direct cost saving. Life cycle costing can reveal such savings on a longer time basis. Or again landscape management to protect habitats may yield development gains through more ready sales or higher prices and rents. Environmentally-friendly properties and developments can capture a niche of their own in a competitive market. And there are health and safety at work issues relating to environmental impacts, as Barker-Read and Farnfield's paper shows.

But while one can search for and sometimes find financial benefits in present value terms from adopting an environmental perspective, it is important also to recognise that such a perspective will probably carry costs. It is not so much that the development or property will require additional investment; even if such investment yields surplus returns, there will be an added cost to the professional's fee. More professional time will probably be spent in assessing the situation and advising on it; more environmental information will need to be collected and considered; the range of options to be considered has now been extended beyond conventional practice. All this imposes costs which have to be borne by client or the professional. Whether the client is willing to bear this cost is the question.

Some clients, particularly the larger ones, may be willing and able to bear the costs. They may take a longer time scale and prefer to avoid the risks associated with ignoring the environment, risks of unforeseen flooding, landslide or contamination, or risks of increasingly stringent environmental regulation. Larger clients may be more vulnerable to the adverse publicity of being seen to be environmentally unfriendly. They may already be aware of the growing climate of claims for negligence and environmental liability. They may prefer to compete with others on the level playing field of increased environmental regulation which is built into anyone's decision making. But smaller clients and those in less stable situations may continue to adopt a short term perspective and see the added costs of environmental investigation and consideration as a net burden. The client mix can therefore constrain the demand for a greener professional.

These are ways in which demand for the land and property manager's environmental skills is affected by the individual characteristics of the client and property or project involved. There is also the issue of how taking environmental considerations on board is liable to affect the property market as a whole or sectors within it, with consequent knock-on effects on the demand for professional services. A recent controversial example was the threatened effect of the proposed registers of contaminated land. Here more environmental information was seen as potentially freezing active exchange in extensive local property markets. New research by Dixon and Richards on valuation of contaminated land also considers this. Similarly the research by Bhatti and Sarno reported here has indicated that providing information on energy rating of domestic property could depress part of the housing market. In these circumstances it is not a straightforward matter to expect professionals to embrace greater environmental awareness. Government regulation may be necessary to get changed practice in those areas where a slow down in the property market and a reduction in demand for professional activities may be a consequence at least in the short term.

But the market for land and property management skills is not simply determined by demand derived from the property market. It is clear that the current condition of apparent oversupply of professionals is also a relevant factor. In a situation of ready availability of services, competition may encourage professionals to provide those services cheaply and this does not encourage investment in expensive environmental services. It does not seem that a wholesale reduction in the output of new managers is the answer here. Rather it would seem that some form of socialisation of part of the costs of providing those services is needed.

Here the education institutions could play a part with some universities providing a specialist form of environmental professional, catering for a high value niche in the market where clients are willing to pay for a service and where the quality of that service is distinguished from the mainstream. This unfortunately does not hold out the prospect of a general greening of the property market. Therefore another strategy would be for The Royal Institution of Chartered Surveyors and other professional bodies, together with governmental support and the involvement of educational and research institutions, to undertake to provide some of the core information needed for environmental assessment and appraisal in the property market. Taking responsibility for some of the costs of such information provision would enable more professionals across the property market to incorporate it in their everyday advice and decisions. This might include underwriting the development costs for specialist software to extend energy rating or life cycle costing. It might involve collation of data on specific market transactions with an environmental dimension. It might involve access to a databank of information on policy, legislation, and instruments, both national and European and even international. It might mean subsidised expertise on understanding environmental impacts and devising means of mitigating those impacts.

THE BROADER CONTEXT OF PROFESSIONAL KNOWLEDGE

So while one can see how the value base of land and property management towards environmental protection can be reinforced and the ways that the market for professional skills may encourage the incorporation of an environmental perspective, one can also see constraints on and barriers to such change. It is therefore instructive to look at the broader context within which professionals must now operate. Many commentators have noted that society appears to be moving into quite a distinctive phase in the late 20th century and into the 21st century (Giddens 1990, Harvey 1989, Beck et al. 1995). In this phase of late modernity or post-modernity as it has variously been termed, specialist knowledge and expertise are frequently challenged. We think we are living in riskier times, including environmentally riskier times, but we are all too aware that knowledge of those risks and how to deal with them is itself uncertain. The notion of an expert providing a clear path towards a better world using scientific knowledge and expecting deference and compliance from an uninformed public is now increasingly challenged. The public, both in general and as clients of professionals, bring much knowledge to bear on their situations themselves. They also expect to have a say in decisions affecting their future. This is a generalisation about professional practice today but it must affect land and property managers as it does doctors, or planners, or teachers.

Therefore a key point about environmental issues and land and property management, one emphasised by several of the researchers, is that professionals must know the limits of their own competence and respect the environmental knowledge of others. In the environmental area, the land and property manager is necessarily working with others: other professionals; the client who possesses much pertinent knowledge; and other groups; local environmental groups themselves can often collect relevant environmental data as with local ecological surveys. This means that the new environmental agenda makes very specific demands on land and property managers. It requires them to learn and refine skills concerned with managing teams of expertise, with networking across many different types of organisation, with conceptual boundaries wherever land and property is concerned. These are not techniques and issues that land and property managers have traditionally concerned themselves with, acting rather as the individual expert. But the environmental agenda is pushing land and property management into recognising not only the rights and needs of future generations but also the changing expectations of professional practice in contemporary society. This is therefore a double challenge for land and property management.

THE ROLE OF THE ROYAL INSTITUTION OF CHARTERED SURVEYORS

In this scenario of change - a new environmental agenda, changing market pressures and a profoundly altered social context - the key British professional body of land and property management, the RICS, is being called upon to respond to and even manage some of these changes. Many of the individual research projects reported here have concluded with specific recommendations for RICS action, rather than say for more governmental action. It would seem that there are six areas of action that a professional body such as the RICS could engage in.

First, as the central professional body of surveying, the RICS can reinforce those elements within the surveyor's value system which stress the protection of the environment and the stewardship of its assets for the future. The statements of the President, the tenor of institutional policy statements, the priorities of the various committees, panels, etc. within the Institution's organisational structure can all reflect and promote such values.

Second, there is the general recognition that more information needs to be provided on environmental issues and that the RICS could take a lead in providing such information and socialising some of the costs of analysing environmental impacts. The research programme reported here is a move towards such a role.

Third, the involvement of the RICS in the education and training of surveyors provides an opportunity for encouraging new generations of surveyors to be highly environmentally aware and specifically trained. The current role of the Institution centres on recognising and reviewing courses run by universities and, therefore, greening the surveying curriculum will involve a partnership between the universities and the Institution. Discussion between these organisations needs to consider the extent to which general environmental education across the board is needed or the training of selected groups in more specialised environmental skills, along with many other issues raised in Parsa's paper. The first task of the RICS must be to facilitate this discussion and to this end the Institution is currently working on a set of definitions of the core knowledge, skills and understanding that chartered surveyors should possess.

Fourth, the RICS has to consider the operation of its central function of the regulation of professional practice in relation to enhanced environmental protection. The enforcement of the new Red Book is a key issue here as is the operation of the professional indemnity scheme. This 'sharp end' of professional regulation may prove to be the key impetus to getting surveyors to take environmental considerations seriously in everyday practice. The RICS can therefore hasten the pace of change in professional practice by sending out the right signals on regulation and enforcement.

Fifth, there is the ongoing lobbying of government that the RICS undertakes as the representative body of the surveying profession. As such, the Institution is frequently asked to comment on policy. In doing so the Institution is able to press for changes which would protect the environment and promote more environmentally responsible land and property management. This is clearly not the only goal that the Institution has in mind in the advice and views it offers to government. And there may well be conflicts between presenting a sustainable development perspective and protecting the interests of property owning clients of the profession. Such conflicts need to be worked out within the Institution if the commitment to stewardship is to remain meaningful when policy is at stake. In making comments on policy, research of the type presented in this volume is clearly invaluable, particularly those where lessons from abroad can be learned, as with Taussik's work comparing English and Swedish coastal zone management.

Sixth, and finally, the RICS needs to look to its own house. It is difficult to legitimately press others, whether clients, educational institutions or government, to take environmental responsibility more seriously if the Institution does not do so itself. It is therefore most encouraging that in 1995 the Executive Group of the RICS has decided that the Institution should have a comprehensive environmental strategy, comprising policies in respect of each area of its Mission Statement and including an assessment of how the Institution manages its own premises. The intention is that this strategy should be in place by July 1996.

The new environmental agenda of the late 20th century poses a great challenge to land and property management and to the associated professions. And it is not an easy challenge. There are fundamental conflicts involved including conflicts between the present and the future. But there is an opportunity for the land and property management profession to take a lead in promoting environmental protection and moving towards sustainable development through changes in land and property management practices. Through a combination of education, information provision and changed professional practice a significant change could occur which will make the surveyor's commitment to stewardship a reality into the 21st century.

REFERENCES

Beck, U., Giddens, A. & Lash, S. (1995). *Reflexive Modernisation*. Polity, Oxford.
Burgess, J. (1990). The production and consumption of environmental meanings in the mass media: a research agenda for the 1990s. *Transactions of the Institute of British Geographers*, **15**,2,136-61.
Daly, H. (1992). *Steady-state Economics* 2nd ed. Earthscan, London.
Ekins, P. & Max-Neef, M. (1992). *Real-life Economics*. Routledge, London.
Giddens, A. (1990). *The Consequences of Modernity*. Polity, Oxford.

Grubb, M., Koch, M., Munson, A., Sullivan, F. & Thomson, K. (1993). *The Earth Summit Agreements: a guide and assessment.* Earthscan with RIIA, London.

Harvey, D. (1989). The Condition of Post-modernity. *Global Environmental Change,* **1**, 3, 183-200. Blackwell, Oxford.

Intergovernmental Panel on Climate Change (1990). *Policy Makers Summary on the Scientific Assessment of Climate Change.* WMO/UNEP, Nairobi.

Kempton, W. (1991). Lay perspectives on global climate change. *Global Environmental Change.* **1**, 3, 183-208. Blackwell, Oxford.

Lovelock, J. (1983). *The Ages of Gaia: a biography of our living earth.* OUP, Oxford.

MacDonald, K. (1995). *The Sociology of the Professions.* Sage, London.

Pearce, D. (1989). *Blueprint for a Green Economy.* Earthscan, London.

Pearce, D. (1995). *Blueprint 4: capturing global environmental value.* Earthscan, London.

Schumacher, F. (1973). *Small is Beautiful: economics as if people really mattered.* Abacus, London.

Worcester, R. (1993). Societal values and attitudes to human dimensions of global environmental change. Paper to *International Conference on Social Values,* Complutense University of Madrid 28 September.

World Commission on Environment and Development (1987). *Our Common Future.* OUP, Oxford.

2 Sustainable Land Management and Development: The Role of the Rural Chartered Surveyor

SUE MARKWELL & NEIL RAVENSCROFT
University of Reading, Reading, UK

INTRODUCTION

Despite the growing intensity of the debate about the relationship between economic development and the continued well-being of the environment, it is only recently that chartered surveyors in the UK appear to have become aware of its significance and potential. Indeed, The Royal Institution of Chartered Surveyors (RICS) makes the point that many chartered surveyors believe that '... there is nothing in the environment for them' (RICS, 1995, 9). In contrast, however, the RICS's rural practice client needs survey (Consensus Research International, 1994) found that nearly half of all clients wanted more advice on environmental matters, but did not expect to gain it from chartered surveyors.

As a prerequisite to undertaking environmental work, therefore, it is clear that chartered surveyors have to develop both new skills and, significantly, a new perception of their work. There is as yet, however, little emphasis being given to the environment as anything other than an emerging market surface. Indeed, rather than clarify the issues, the recent guidance note on environmental management (RICS, 1995) rather compounds them. Comprising six individual papers, it represents a step in developing an agenda for the role of chartered surveyors in the environmental management and development of land and property. However, this agenda is not without tensions. In particular, there is a fundamental tension between the strategic construction of the environment as a basis for realigning corporate policy and management (Heywood, 1995 and Hymas et al., 1995) and the more limited evocation of environmental issues as potential new business (Jayne, 1995). This tension is redolent of a wider debate over the competing environmental narratives of ecology and commodification (Whatmore and Boucher, 1993). While both narratives construct the environment as an outcome of socio-economic processes, the former articulates it as possessing intrinsic value, while the latter ascribes to it no more than the instrumental value

The Environmental Impact of Land and Property Management, Edited by Yvonne Rydin
© 1996, The Royal Institution of Chartered Surveyors

of a marketable cultural artefact (Hattingh et al., 1994, Murdoch, 1993 and Whatmore and Boucher, 1993).

Faced with such contradictory guidance, it is little wonder that rural surveyors display scepticism about the pertinence of the environment to their work. As a result, they remain for the most part interested in the environment only insofar as it offers market opportunities or, increasingly, reduces their liability for negligence (Hymas et al., 1995). There is, therefore, little indication of their current level of environmental knowledge. Nor is there any indication of a widespread commitment to sustainable development. Consequently, there is scant evidence of the development of environmental policies suitable to the work of rural surveyors, meaning that little is known about how such policies might alter working practices or how they might impinge on the delivery of professional and managerial services.

This paper will seek to address these issues, first by discussing the nature of sustainable development and then progressing to assess the extent to which it is relevant to the business of rural surveyors. Based upon this work, the paper then reports the findings of a survey on the current attitudes and practices of rural practice chartered surveyors before developing some principles of good practice capable of general application in the rural profession.

SUSTAINABLE DEVELOPMENT AND THE ENVIRONMENT

From the publication of the Brundtland Report in 1987, the concept of sustainable development, that is: '..(*development that meets the needs of the present without compromising the ability of future generations to meet their own needs'* (World Commission on Environment and Development, 1987, 43) has become an increasingly central part of global environmental strategy. As Gorz (1987) argues, prior to Brundtland questions of resource conservation had effectively been by-passed by the Western imperative of 'environmentalism through growth'. Thus rather than addressing the essential problematic of human 'progress', most concern had been focused on the environmental constraints to that progress. In the wake of Brundtland it is now clear that sustainable development must involve changes in both consumption patterns and in investment patterns, to augment rather than deplete environmental capital (Pearce, et al., 1989).

Rather than act as the catalyst for the necessary paradigm shift, however, the concept of sustainable development has effectively been appropriated as a means of reconciling the apparent differences between development and the environment (O'Riordan, 1988 and Murdoch, 1993). This is consistent with the neo-classical 'logic' of the constant pursuit of growth, with its inherent creation of dilemmas rather than solutions (Gorz, 1987). Indeed, one of the enduring 'tricks' of neo-classical discourse has been the elevation of 'public goods', such as the environment, to a position of prominence beyond the market, known metaphorically as 'market failure' (see Pearce, 1993). Having achieved this moral 'high ground' of depicting the technical failure of the market, the discourse then proceeds to reconstruct the failure to produce a socially efficient allocation of resources as the failure of society to attribute sufficient value to such goods (Voges, 1994).

As a result, it is increasingly apparent that rather than precipitating the paradigm shift, sustainable development actually suggests a 'comfortable reconciliation' with the dominant Western ideology of industrialisation (Torgerson, 1995). This inevitably leads to the position where the very ambiguity of the logic is exploited to ensure the continuation of the dominant discourse. Rather than assuming its position as a limit to growth, therefore, sustainable development actually becomes the very reason for growth. Using the principles of substitutability and net environmental gain, development can be framed as the primary means through which wealth creation can simultaneously increase the overall stock of environmental resources.

The Brundtland Report identified the twin concepts of need and technological limitation as being at the heart of sustainable development. This has since been questioned by Pearce et al. (1993) who suggest that sustainable development rests on the dual assumption that current generations are both adversely and unfairly affecting the well-being of future generations. The first element of this assumption is, according to Pearce et al. (1993), largely a question of fact, leaving the essential issue as a question of the morality of intergenerational and, by inference at least, temporal intragenerational equity.

This issue of morality has thus become a central theme in the work of resource managers such as rural surveyors leading, suggest Hattingh et al. (1994), to them assuming a range of 'value-positions' with respect to the environment. These positions are termed: Egoist; Stewardship; and Unity with Nature. While the first of these implicitly rejects any notion of futurity, the remaining two attempt to deal with it according to, respectively, anthropocentric and ecocentric principles. While the 'Unity with Nature' scenario is the only one to attribute an inherent value to the environment, it also eschews any form of development. This leaves the utilitarian 'Stewardship' option as the most suitable for balancing the environment and economic development, even with its instrumental approach to environmental value.

Similarly, Pearce et al. (1993) have developed a 'sustainability spectrum' (see Table 1) to represent the range of responses to sustainable development. The range covers both technocentric and ecocentric responses, suggesting a four-part division from least to most sustainable. At its extremities, the sustainability spectrum bears a strong resemblance to Hattingh et al.'s (1994) typology of value-positions, in labelling the Egoist, or in Pearcean terms Cornucopian, response as unsustainable, in contrast to the Unity with Nature, or Deep Ecology, response as very strongly sustainable.

Rather than these extremes, it is suggested that most current attitudes fall in the central zones, of 'weak' and 'strong' sustainability. Weak sustainability,

Table 1 Different perspectives on the sustainability spectrum

Pearce et al. Sustainability Spectrum	Cornucopia	Accommodating	Communalist	Deep Ecology
	very weakly sustainable		**very strongly sustainable**	

TECHNOCENTRIC   ECOCENTRIC

←——→

	environment is valueless		***environment has inherent value***	
Hattingh et al. value positions	Egoist	Stewardship		Unity with Nature
Welford's SME characterisation	Ostriches/ Laggards	Thinkers		Doers

characterised as 'Accommodating', displays many of the central attributes of the Hattingh stewardship model, being based upon a form of anthropocentric conservationism overridingly concerned with intra- and intergenerational equity. This, Pearce, et al. (1993) argue, is the most favourable construction of the outcome of current policies. In contrast, they argue, global policies should be seeking to encourage a shift towards strong sustainability, characterised as 'Communalist'. Under this scenario the economy shifts from a 'green market' perspective to a highly-regulated steady state in which the interests of the collective take precedence over those of the individual. In calling for this shift, Pearce, et al. recognise that:

> *'The notion of a "sustainable" society is radical. Sustainable development confronts modern society at the heart of its purpose, because the human race is, and always has been a colonising species without an intellectual or institutional capacity for equilibrium (1993, 184).'*

RURAL PRACTICE SURVEYORS AND SUSTAINABLE DEVELOPMENT

The issues facing rural chartered surveyors are, therefore, at the very heart of their purpose as resource professionals. However, while some progress is being made in relating sustainable development to business management, it remains far from clear how it relates to service industries such as surveying. What is certainly apparent, from sources such as the RICS guidance note and client needs survey, is that rural surveyors are currently operating in a form of environmental 'vacuum' in which the consumerist imperative of 'environment as market opportunity' has come to dominate any form of ecological standpoint.

In order to establish how far rural practice surveying firms have taken on an environmental perspective in their work a questionnaire survey, derived from the work of Welford and Gouldson (1993) and Elkington (1987), was conducted on a stratified 5% sample of rural practice chartered surveyors working in private practice, land agency and the voluntary, public and corporate sectors.

While seeking a broad range of information on environmental practice and opinions, the survey found relatively little evidence of action on sustainable development outside of the public and corporate sectors. Instead, the current emphasis is very much on the quality of services, with nearly one-fifth of the offices, mainly in the private sector, now having, or working towards, a quality management system approved under BS5750 (EN ISO 9000). Regardless of this, the respondents claimed overwhelmingly that environmental issues **were** significant to their businesses, with approximately 90% claiming them to be 'important' or 'very important'.

Most respondents claimed that they were already undertaking 'environmental' work. The type of work cited as being 'environmental' was, however, very wide-ranging, with the most commonly undertaken being habitat protection, health and safety issues, planning applications, land use issues and pollution control. Relatively few respondents undertake work in the more specialised areas of environmental auditing or policy, contaminated land or climate-related work, while work in areas such as energy conservation appears to be of increasing significance (see Figure 1).

The growing significance of this environmental work to rural practice chartered surveyors is probably most clearly evident from the level of turnover associated with it. While public sector employees suggest that environmental work may now account for about half of their gross income, the private sector indicates that it is now accounting for approximately one-fifth of turnover. However, in contrast to this background of optimism over the financial potential of environmental work, just 35% of respondents claim that they keep up-to-date with current environmental legislation, with a further 30% claiming to be 'mostly' up-to-date. This contrasts with nearly 40% of those currently undertaking some elements of environmental work who admit to being less than fully informed about relevant legislation.

In terms of sources of information, it is clear that the RICS is currently filling a vital role as the major conduit of relevant material (see Figure 2). Apart from those in the corporate sector, who tend to use legal journals as their primary source of information, and those in the public sector, who use much government material, all the other categories of surveyor look to the RICS as their chief source of information about environmental issues. With the publication of its guidance

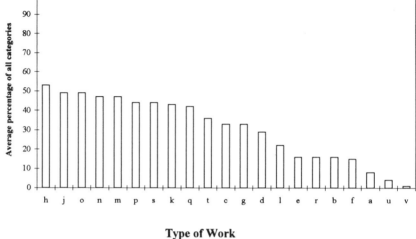

Type of Work

h=habitat	j=health /safety	o=planning applications
n=pollution control	m=land management	p=housing
s=visual amenity	k=heritage/leisure	q=timber
t=waste control	c=energy	g=ground water
d=environmental audits	l=land consultancy	e=environmental law
r=transport	b=contaminated land	f=environmental statements
a=climate	u=mining	v=land fill

Figure 1 Type of environmental work undertaken

notes on 'Environmental Management and the Chartered Surveyor' the RICS has clearly taken a step in the right direction in contributing to the environmental awareness of land managers.

Approximately one-quarter of the respondents claim to have developed environmental policies, the majority written. However, if the public and corporate sectors are excluded, just 10% of respondents work in private organisations with an environmental policy, while in only 4% of cases is this a written policy. Equally, with the exception of the public sector, where environmental policies have existed for a number of years, few policies have existed for more than five years.

For the most part, environmental policies have been drawn up by senior management, although external consultants have been involved in about one-third of the cases. Senior management is similarly largely responsible for the monitoring of environmental performance, with external consultants being monitoring of environmental performance, with external consultants being involved in less than 20% of cases. Moral judgements are expressed as the principal reason for developing environmental policies but alongside this motivation has been a belief by some that there are public relations benefits to client pressures, either in a positive sense, or in an attempt to limit their possible

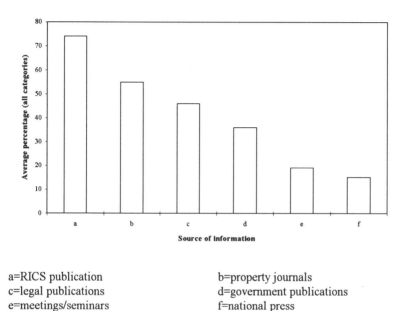

a=RICS publication   b=property journals
c=legal publications   d=government publications
e=meetings/seminars   f=national press

Figure 2 Sources of information on environmental issues

future liability for negligence (see Figure 3). Indeed, negligence is a major factor for private practice whereas for land agents the morality of deriving an environmental policy has overwhelmed other considerations. The corporate sector, for the most part, views the implementation of an environmental policy as part of its public relations exercise.

Although the coverage of the environmental policies is variable, the most common areas dealt with are health and safety, energy conservation, staff training, waste management and general environmental awareness (see Figure 4). While considering energy conservation an important issue, less than half of the environmental policies (and very few at all in the private sector) deal, for example, with policies for staff transport. In addition, issues such as pollution control, land use implications and the social impacts of environmental management are not common elements of environmental policies for firms, although it is in these areas that chartered surveyors are most active in advising features very strongly in the private sector, while health and safety, pollution control and land use issues are dominant in the public sector.

Only in the case of the corporate sector are environmental policies applied universally, to internal operations, the management portfolio and to suppliers. For the remainder, application is split fairly equally between internal, focusing on the operation of the business, and external, relating to professional advice, with

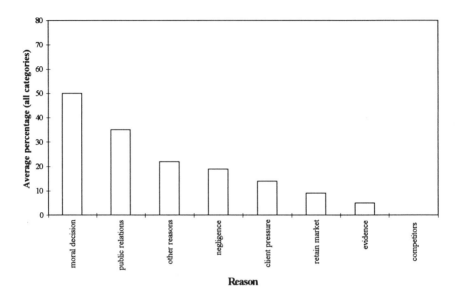

Figure 3 Reasons for drawing up an environmental policy

relatively little attention paid to suppliers. Thus, while the majority of public sector environmental policies guide internal operations, over half of them relate to professional work or portfolio management and about one-third to their suppliers. Conversely, three-quarters of private sector environmental policies cover professional work for clients, while less than half relate to internal organisation and very few to suppliers.

While these results suggest that there is interest in, and activity associated with, environmental policies, evidence on monitoring and staff training indicates that it may be largely superficial. While less than one-fifth of policies are monitored on a regular basis (and only one-quarter of these externally to the organisation), around one-quarter of the policies are subjected to no apparent monitoring at all. Of the remainder, the majority are subjected to what is commonly described as 'periodic internal' review only. Similarly, in few cases are all staff trained in environmental management, while in less than half of the offices are all of the professional staff given appropriate training.

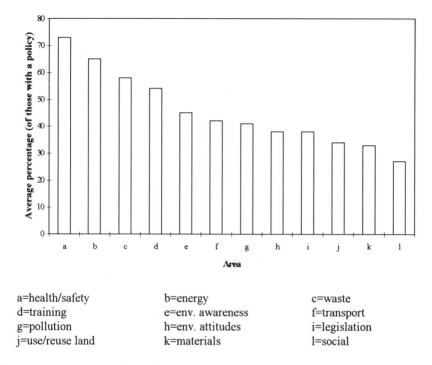

a=health/safety   b=energy   c=waste
d=training   e=env. awareness   f=transport
g=pollution   h=env. attitudes   i=legislation
j=use/reuse land   k=materials   l=social

Figure 4 Areas covered by environmental policies

Of those businesses which have not developed an environmental policy, the overwhelming reasons for not doing so are that they are not considered relevant or necessary to the business (see Figure 5), with respondents making comments such as:

> *'I don't understand why there should be a need for a formal environmental policy for our business (private sector).'*

and:

> *'There is not a perceived need for such a policy (private sector).'*

The results of the survey therefore indicate an apparent polarity in rural surveying, where environmentalism either assumes some significance, or it does not. Lack of appropriate policies is therefore less a function of size, cost or expertise than it is a signal of the primary focus of the business, although some respondents did suggest that a lack of time and resources had prevented them from formalising their policy. This is borne out by the reasons given for developing an environmental policy in the future. Rather than the moral imperative evident in those already operating environmental policies, the emphasis is on opportunism and expediency, particularly in terms of client pressure, the possibility of new business opportunities or warding off competition.

It is thus apparent that the overall position is one in which the majority of businesses currently pay little more than lip-service to environmental issues. Even in cases where environmental policies have been developed, they are often inconsistent and are likely to be subject to inadequate monitoring or operation by inadequately trained staff. However, it is important to recognise how recently sustainable development became an issue and, consequently, how little guidance currently exists. As a result, much of the apparent opportunism may be more a reflection of a learning curve in which the current level of discourse is still highly market driven.

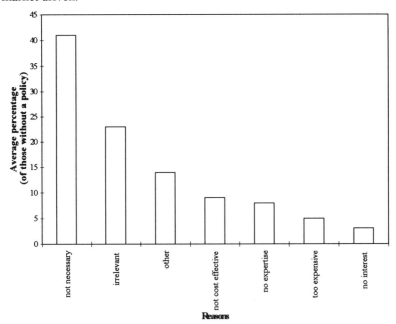

Figure 5 Reasons for not having a policy

Perhaps reflecting this learning curve, the individual respondents appear much more highly environmentally aware than the business responses would suggest. In the majority of cases they are already actively pursuing environmental policies at home. Often this involves little more than recycling waste, although nearly half consider the environmental implications of their food purchases, while over 30% have made explicit environmental decisions about their personal transport.

It is therefore clear that the chartered surveyors interviewed can readily separate their own views and values from those of their businesses. While apparently actively modifying their home lives in an environmentally-friendly manner, few respondents seemed to carry this through to their business life. This corresponds with the findings of Welford, with respect to the managers of small businesses:

> '*Clearly, on a personal level they are keen to see environmental improvement but they are less keen to see their businesses pay for this(1994, 159-160).* '

Beyond the legal imperatives of health and safety legislation, therefore, relatively few businesses apparently accord a high priority to issues such as waste control, energy conservation or green transport policies. Rather, most emphasis is clearly given to the market opportunities brought about by the environmental concerns of others, particularly where these become the subject of grant aid, as is often the case with habitat restoration, pollution control and visual amenity.

As a consequence, very few of those responding to the questionnaire conform to anything other than the Pearcean construct of Cornucopian technocentricity (see Figure 1). While the high levels of recognition of environmental issues tend to indicate at least a superficial acceptance of the environmental imperative, the rhetoric is rarely matched by the apparent actions. Environmental responsibility is, as Shrivastava (1995) has recently suggested, still located at the margins of business, with the core decisions on finance, markets and clients still largely unaffected or unmodified. What emerges from this is really the opposite of the holistic notion of environmental management. Rather than a comprehensive appraisal of the relationship between rural practice surveying businesses and the environment, the results indicate a high degree of segmentation and opportunism.

There are, of course, exceptions to this, with some businesses clearly attempting to come to grips with the implications of environmentalism. Approximately one-fifth of the respondents now operate a quality management system under BS5750. This is dominated by private practice, where 34% of the businesses surveyed have accreditation. While only one respondent, in the corporate sector, is actively pursuing accreditation under BS7750, a further 20 businesses have an environmental policy, mostly derived at the corporate level. It remains, however, that few firms yet have a comprehensive approach involving both their own businesses and their advice given to clients.

As a prerequisite to rural surveyors accepting more fully the tenets of sustainable development, it is important to assess their potential for improvement. Very little work has been done on this area (Lowe and Murdoch, 1993), with the most applicable probably being that of Welford (1994) and Welford and Gouldson (1993) on the small and medium sized enterprise (SME) sector of the British economy. In many ways the SME sector of the economy has much in common with chartered surveying businesses. Taken individually, their impact on the environment may be small, but in overall terms the impact could be extremely significant, especially in the case of the land area of Britain managed by chartered surveyors. However, the size of individual enterprises and businesses often limits or precludes them from undertaking sophisticated environmental reviews or adopting rigorous environmental policies.

Evidence suggests, however, that rather than develop simple principles or guidelines, the majority of those in the SME sector show little interest in the environment at all. A recent survey of the key impediments to implementing environmental strategies or improvements (Welford, 1994) confirms this position. Whereas in 1992 the primary disincentive to adopting environmental measures was their perceived lack of relative importance, by 1993 their importance was increasing, but not to the point where their financial implications were considered cost-effective. Similarly, while surveying firms are willing to advise clients on a wide range of environmental issues, from habitat protection and health and safety measures to pollution and waste control, few chartered surveyors have yet seen the need to extend this environmental concern to their own business practices. Indeed, the survey indicated that many chartered surveyors perceive good environmental management to be no more than a case of applying 'common sense' to the management of their clients' land:

> *'Although we have no formal policy, employer/employees have a fairly sharp awareness of environmental factors and we run the estate's business on a common sense/responsible basis.'*
> *(land agent interview)*

From the analysis of his survey, Welford (1994) has produced a four-part characterisation of the SME sector which correlates with Pearce et al.'s (1993) sustainability spectrum (Figure 1). Rather than representing the full scope of the spectrum, Welford places three of his four groups firmly in the unsustainable 'technocentric' portion of it. Indeed, about three-quarters of Welford's sample (described as 'ostriches' and 'laggards') belong to the Cornucopian extreme of technocentricity, where they choose not to recognise the environmental challenge in their continued exploitation of resources.

As the largest group, the 'ostriches' appear to believe that environmentalism is a passing phase, while the 'laggards' claim to recognise the challenge while being unable to act (Welford, 1994). These two characterisations are strongly redolent of the majority of views of those interviewed in the survey of rural surveyors. Indeed, it would seem to place most surveyors in the 'laggards' category, with their rhetorical recognition of the environmental imperative matched only by their lack of action in tackling it.

Of those few surveyors going beyond rhetoric, all belong in Welford's next category, of 'thinkers'. Still firmly technocentric, thinkers belong to the 'accommodating' category, with its emphasis on a managerialist response to the sustainability imperative. None of those interviewed could be termed 'doers', Welford's fourth but only ecocentric category.

ENVIRONMENTAL STRATEGIES AND THE OPERATION OF SURVEYING BUSINESSES

In overall terms, therefore, it is apparent that the current environmental performance of surveyors, in common with the whole of the SME sector, falls some way short of that required for the integrated implementation of sustainable development policies. In considering how this might be rectified, Welford and Gouldson (1993) suggest attention to the four key components of an environmental management system: review; policy; design and implementation of the system; and auditing. This corresponds with the requirements of BS7750, Environmental Management Systems, where the emphasis is very much on the development of a system capable of meeting its own objectives.

In deriving an environmental management system therefore, the first stage is the baseline review of the current environmental performance of the firm. As an example, the LEAF (Linking Environment And Farming) environmental audit system (Table 2) used by Grays Management puts much emphasis on participants recording their own current performance as the basis for future annual reviews.

Although concentrating very much on arable farming and therefore of limited use to the businesses of many chartered surveyors, the LEAF audit system has proved of use to Grays Management, which reports that the benefits include helping to identify areas of a farm or estate which require special attention, helping to raise the standard of estate management, and improving the financial performance of the company.

Once the baseline review has been completed, the firm can proceed to consider its environmental policy. As the firm's statement of objectives, the policy must outline clearly the firm's commitment to environmental improvement (Welford and Gouldson, 1993). Equally, the policy must be sufficiently detailed to allow the definition of both future actions and their measurement. One such approach to environmental policy has been developed by the Environment Challenge Group (MacGillivray, 1995) as a set of indicators, many of which are highly pertinent to chartered surveying businesses. These indicators can be broken down further into those that are pertinent to the business and those that relate to clients (Table 3).

Table 3 illustrates clearly the interrelationship between the management of a surveying firm and the advice given to clients. Many categories of advice are similar, implying that client advice should be capable of application to the business. However, the more important concern is the ability to identify significant issues and their associated implications. Recognition of how the management and use of a client's land might affect the climate thus implies the concurrent recognition of how the surveyor's business might do likewise, even if at a less significant level.

Table 2 Linking environment and farming

Linking Environment And Farming

The LEAF Environmental Audit provides all farmers with an opportunity to record and monitor, on an annual basis, their own environmental policy in the context of an economically viable business' (LEAF publicity leaflet).

The objectives of LEAF are:

* to develop the concept of Integrated Crop Management (ICM) as being synonymous with responsible farming;
* to demonstrate that through ICM farmers can produce a consistent economic yield of high quality food at reasonable cost;
* to demonstrate that the responsible use of crop production inputs is necessary and does not lead to environmental degradation or health risks;
* to promote ICM to farmers and its benefits to the public.

The LEAF Audit involves annual self-assessment in seven areas of operation:

* wildlife habitats and landscape;
* soil management;
* crop protection;
* energy usage;
* pollution control;
* organisation and planning;
* animal welfare.

Source: LEAF Annual Review (1992/93)

Once in place, an environmental policy must be backed up with an appropriate process of implementation, such as an action plan to cover the specific objectives necessary to fulfil the environmental policy. Two of the crucial issues involved in implementing an action plan are the identifying of appropriate staff and the allocation of sufficient resources to allow them to carry out their tasks. A good illustration of staff training is provided by English Nature's environmental management policy, contained in its recently published booklet 'A Guide to Environmental Management' (1995), in which the organisation details its initiatives in taking on specific responsibilities to reduce energy consumption, to conserve natural resources and to minimise waste and pollution by working to a detailed action plan (Table 4).

Although the last stage in the process of environmental management, reporting assumes considerable importance, since it is only really at this stage that the full impact of the strategy will become clear. Yet it is at this point that most of the survey respondents failed to capitalise on their efforts, either through inadequate

Table 3 Environmental Indicators

BUSINESS-RELATED

* the effect of its activities on the climate,
  (e.g. CFC emissions, air pollution from office machinery)
* use and reuse of land
  (e.g. location of office premises and landscaping of immediate area)
* internal energy management
  (e.g. heating and air conditioning systems, double glazing and heat loss prevention)
* selection of materials for office buildings
* effects of business on culture and heritage
  (e.g. use of local materials in building, local workforce and suppliers)
* transport policy
  (e.g. use of public transport, car-sharing, fuel-efficient cars)
* environmental assessments and audits of the business
* raising environmental awareness in staff
  (e.g. staff training)
* improving environmental attitudes of suppliers
  (e.g. through explicit purchasing policies)
* compliance with regulations and legislation

CLIENT-RELATED

* effects of management of clients' land on climate
  (e.g. large-scale felling of timber)
* the efficient use of the land resource
  (e.g. possible use of LEAF audits)
* advice about energy management for buildings and machinery
* advice on design and construction of buildings
* the protection of heritage and culture
* transport policy for clients' business
* offering environmental assessments and audits of clients' property and business
* raising environmental awareness of client and his employees
* control of pollution on, over and under the client's land
* advice on management of hazardous products used on clients' land
* advice on compliance with current legislation
  (e.g. in relation to pollution control)

Source: Derived from MacGillivray (1995)

Table 4: English Nature Plan for Environmental Management

English Nature seeks to:
* address all environmental management issues, especially where there are implications for nature conservation;
* maintain an awareness of relevant environmental legislation and ensure our full compliance with it;
* set objectives for sound energy management through monitoring and targeting and efficient use of resources;
* minimise the use of all materials through re-use or recycling;
* match our internal practice with our policy statements;
* help our contractors and suppliers to recognise their environmental responsibilities;
* promote environmental management awareness amongst staff; and
* publicise our environmental performance and achievements in Annual Reports.
Source: English Nature (n.d.)

monitoring or reporting, or no monitoring or reporting at all. Indeed, formal external reporting appears to be largely limited to public organisations with statutory reporting responsibilities. An example of this is the Countryside Commission's Annual Report (see, for example, Countryside Commission, 1994), which is laid before Parliament in accordance with the Wildlife and Countryside Act 1981.

Similarly, incorporated in the English Nature Action Plan is the commitment to publicise its environmental performance and achievements as part of its Annual Report. So although there is no explicit reference to performance monitoring, it will have to occur on at least an annual basis to facilitate the reporting function. No reference is made, however, about the extent to which either the performance itself, or the monitoring or reporting is audited, whether internally or externally.

Rather than use the medium of the annual report as the chief conduit for reporting environmental performance, National Power publishes an annual Environmental Performance Review (see National Power PLC, 1994). This records company progress towards specific targets in all key areas of operation, covering reducing emissions, improving resource usage, strengthening environmental management, improving public accountability and developing international projects. As an example, the introduction of BS7750 is being supported as a means of strengthening environmental management, with pilot trials having been completed prior to full implementation. In addition, management plans are being drawn up to BS7750 standard for all National Power's non-operational sites:

'As part of the company's site management plans, provision is made to protect and, where possible, enhance the environment

for the benefit of a range of rare species.'(National Power PLC, 1994, 22).

Alongside this process should be the external monitoring of the evolution of the green agenda, to ensure that the environmental management system adopted by the business remains suitable for its purpose. While this monitoring can be essentially passive, Elkington (1987) suggests that many firms have a more active role to play, both by contributing directly to environmental programmes through forms of corporate sponsorship, and by helping to build bridges between different factional interests.

CONCLUSIONS

That rural practice surveyors have little option but to engage in the environmental debate is becoming increasingly clear, perhaps even as a survival strategy. A refusal to engage will lead not only to surveyors being denied access to a new and expanding market opportunity, but it could also undermine their continuing claim to be 'the property profession'. Moreover, environmental awareness in business is increasingly being viewed as a question of ethics, effectively calling into question any organisation which does not consider the environment as an integral element of its corporate strategy (see McCloskey and Smith, 1995, and Roberts, 1995).

However, the overriding response of rural practice chartered surveyors has been one of indifference and apathy. While there is a fair level of environmental awareness and activity, much of this is either partial or superficial or, predominantly, both. Even where there is activity it often involves no more than appropriating the name of environmentalism to sundry work practices or 'innovations' in client services.

Undoubtedly, the ability of surveyors to embrace the imperative of environmental management is hampered by a lack of knowledge and evidence of good practice, together with a high degree of conflict over the very meaning of the term 'sustainable development' (Pearce, et al., 1993). It is this latter issue which is arguably at the heart of the problem, where sustainable development has been effectively appropriated by the conflicting narratives of managerialism and ecology.

Faced with these problems, it is clear that the central issue for rural practice chartered surveyors is how to endorse the need for a major change in their value systems in a form compatible with, rather than antithetical to, their wider commercial concerns (Roberts, 1995). However, rather than being a wholesale constraint to commercial activity, there is clear evidence from both the survey and the wider literature that, handled correctly, environmental management can improve long-term financial performance. Adopting enhanced business behaviour towards the environment is, therefore, very much an act of faith, based on the extent to which both the profession and its clients can be encouraged to support, accept and act upon new forms of advice and business operation. Ultimately, however, it is apparent that attitudes must change to reflect a new agenda in which the criteria for successful business will be quite different from the present.

It is easy to become overawed by the very size and complexity of the issues to be faced. Many surveyors expressed the opinion that the little they could contribute was irrelevant on the broader scale. Indeed, it is of little surprise that surveyors working in the property division of National Power are more versed in environmental management than those managing traditional estates. Yet, as Roberts stresses, it is easy:

> '... to ignore the possibilities of implementing a succession of small, and in themselves insufficient, actions that cumulatively can bring about the resolution of problems' (1995, 242).

The challenge is therefore clear. Evidence indicates that all businesses have the ability to contribute to environmental improvement, both internally through sound environmental management and externally through the advice they give to their clients. Some businesses are already making advances in incorporating environmentalism within their work patterns, even though the commercial benefits of doing so have really yet to emerge. However, as environmental demands and responsibilities grow, it will be these latter businesses which will be in a position to flourish and prosper.

RECOMMENDATIONS

The recommendations to emerge from this research fall into three categories: action by chartered surveyors; action by The Royal Institution of Chartered Surveyors; action in terms of further research.

Action by chartered surveyors

Integration is the key to environmental strategy for any business. It is therefore necessary for chartered surveying firms not only to examine every aspect of their *own* environmental performance but to look also at the consequences of any advice that is given to their clients. Rural practice chartered surveyors have tended, in the past, to adopt a Cornucopian stance on environmental issues. A shift along the spectrum towards greater sustainability can only be accomplished with a sustained commitment to a fully integrated management policy which combines both internal performance and client advice. The development of an environmental strategy must, therefore, start with real commitment on the part of the whole organisation if it is to be successful, and will necessitate some restructuring of the corporate culture to incorporate the principles underlying sustainable development and environmental management.

It is recommended that rural practice chartered surveyors take steps towards adopting the principles of sustainable development and environmental management laid out in Tables5 and 6 by reviewing, in a holistic sense, their own management strategy alongside advice given to their clients.

Table 5 Principles of sustainable development for rural practice chartered surveyors and their clients

---

Sustainable development is economic development that lasts. It depends upon humans living within the capacity of the planet by ensuring that 'Development meets the needs of the present generation without compromising the ability of future generations to meet their own needs'
(Brundtland 1987).

1   **The protection of critical natural capital**
i.e. natural resources that are critical to well-being or survival and are not substitutable
Action:
* reduction in the use of ozone-damaging materials and global warming, by reducing CFC emissions and use of power sources,
* prevention of the loss of bio-diversity by the sensitive use of land,
* non-use of tropical hardwoods.

2   **The maintenance of natural capital**
i.e. natural resources that are non-renewable and may be non-substitutable
Action:
* minimal use of natural minerals. Use alternatives and recycle materials wherever possible,
* use alternative sources of energy e.g.: wind power for the generation of electricity, and petrol substitutes (bio-fuel) for transport,
* reduction in the level of waste and its careful disposal.

3   **The efficient use of renewable natural capital**
i.e. resources which are renewable and are capable of being replaced by human action
Action:
* reduction in air, land and water pollution,
* efficient use of land and buildings,
* non-depletion of stocks in farming and forestry by careful husbandry and re-stocking.

Action by the RICS

As the professional body in the UK for chartered surveyors, the RICS must take the lead in developing an agenda for sound environmental management. It could best demonstrate its own commitment to sustainable development by the adoption of the principles of environmental management.

It is recommended that the RICS must take steps to develop an environmental policy and management system for its own organisation and must encourage others to follow suit.

Table 6 Principles of environmental management for rural chartered surveyors.

1	Assess the possible environmental impact of all operations and procedures connected with the business by conducting an environmental review.
2	Draw up a suitable and achievable set of objectives, targeted at specific areas, which takes the individuality of the business and its clients into account but which is based upon:
	* minimal use of all materials and supplies and the use of alternative, renewable materials where possible,
	* control of waste and pollution and the use of recycling where possible,
	* efficient use of buildings, transport and energy sources,
3	Encourage the adoption of high environmental standards from all those involved in the business. This includes staff training and raising the general environmental awareness of senior management, employees, suppliers and clients.
4	Strive to be aware of, and to comply with, all applicable legislation and regulations which are designed to improve environmental standards.
5	Monitor the success of the environmental management policy on a regular basis and realign the policy if necessary.
6	Demonstrate your commitment to environmental improvement by reporting your achievements on an annual basis.

In terms of information there is clearly a lack of comprehensive literature which is readily available and in an appropriate form for chartered surveyors. It is also clear that for the majority of respondents in the survey the RICS represents the primary source of their environmental information. Both the lack of information and environmental knowledge require rectification before any major advances can be made in combining environmental management policies into the business practices of chartered surveying firms.

It is further recommended that the RICS takes every opportunity to stress the importance of environmental issues to the members of the profession and to increase the flow of appropriate information on the subject. To this end, it is recommended that the RICS produces a regular 'Environmental Bulletin' giving up-to-date information on environmental issues for all members of the profession.

Action for further research

Action needs to be taken to develop model systems of land management plans which are capable of incorporating the standards appropriate to both BS5750 (Quality Management Systems) and BS7750 (Environmental Management). Evidence exists that work is being carried out in this area, in the private as well as the public and corporate sectors, but that there is at present very little liaison or co-ordination.

It is recommended that research is undertaken to establish how client advice can be framed within the principles of sustainable development and environmental management.

REFERENCES

Consensus Research International (1994). *Rural Property: research into customer needs*. London. The Royal Institution of Chartered Surveyors.

Countryside Commission (1994). *Annual Report 1993-1994*. Publication CCP456. Countryside Commission, Cheltenham.

Elkington, J. (1987). *The Green Capitalists. Industry's search for environmental excellence*. Victor Gollancz Ltd., London.

English Nature (1995). *Guide to Environmental Management*. Peterborough, English Nature.

Gorz, A. (1987). *Ecology as Politics*. (trans. by P. Vigderman and J. Cloud). Pluto Press, London.

Hattingh, J., Voges, I., Muller, K. & Verwoerd, W. (1994). *The relationship between ethics, environment and development: guidelines for policy making in South Africa*. Report prepared for the Development Bank of South Africa. Unit for Environmental Ethics, University of Stellenbosch, South Africa.

Heywood, A. (1995). Corporate environmental statements and the Chartered Surveyor. In The Royal Institution of Chartered Surveyors. *Environmental Management and the Chartered Surveyor*. RICS Guidance Note. RICS Business Services Ltd., London. 29-34.

Hymas, M., Wood, G. & Fox, P. (1995). Environmental audits of land and property. In The Royal Institution of Chartered Surveyors. *Environmental Management and the Chartered Surveyor*. RICS Guidance Note. RICS Business Services Ltd., London. 85-107.

Jayne, M. (1995). The role of the Chartered Surveyor in environmental management. In The Royal Institution of Chartered Surveyors. *Environmental Management and the Chartered Surveyor*. RICS Guidance Note. RICS Business Services Ltd., London. 11-27.

Lowe, P. & Murdoch, J. (1993). *Rural Sustainable Development*. Strategy Review: Topic Paper 1. Rural Development Commission, London.

McCloskey, J. & Smith, D. (1995). Strategic management and business policy-making: bringing in environmental values. In Fischer, F. & Black, M. (eds). 1995. *Greening Environmental Policy: the politics of a sustainable future*. Paul Chapman Publishing Ltd., London. 199-209.

MacGillivray, A. (ed). (1995). *Environmental Measures: indicators for the UK environment*. Environment Challenge Group, London.

Murdoch, J. (1993). Sustainable rural development: towards a research agenda. *Geoforum* 24(3): 225-241.

National Power PLC (1994). *National Power Environmental Performance Review '94*. National Power PLC, Swindon.

O'Riordan, T. (1988). The politics of sustainability. In Turner, R.K. (ed). *Sustainable Environmental Management: Principles and practice*. Belhaven Press, London. 29-50.

Pearce, D. (1993). *Economic Values and the Natural World*. Earthscan Publications Ltd., London.

Pearce, D., Markandya, A. & Barbier, E.B. (1989). *Blueprint for a Green Economy*. Earthscan Publications Ltd., London.

Pearce, D., Turner, R.K., O'Riordan, T., Adger, N., Atkinson, G., Brisson, I., Brown, K., Dubourg, R., Fankhauser, S., Jordan, A., Maddison, D., Moran, D. and Powell, J. (1993). *Blueprint 3: Measuring sustainable development*. Earthscan Publications Ltd., London.

Roberts, P. (1995). *Environmentally Sustainable Business*. Paul Chapman Publishing Ltd., London.

Royal Institution of Chartered Surveyors. (1995). *Environmental management and the Chartered Surveyor*. RICS Guidance Note.. RICS Business Services Ltd., London

Shrivastava, P. (1995). Industrial and environmental crises: rethinking corporate social responsibility. In Fischer, F. & Black, M. (ed). *Greening Environmental Policy: the politics of a sustainable future*. Paul Chapman Publishing Ltd., London. 183-198.

Torgerson, D. (1995). The uncertain quest for sustainability: public discourse and the politics of environmentalism. In Fischer, F. & Black, M. (eds). *Greening Environmental Policy: the politics of a sustainable future*. Paul Chapman Publishing Ltd. London. 3-20.

Voges, I.F. (1994). Environmental management: implementing the paradigm shift. Proceedings of the 19th Annual Conference of the National Association of Environmental Professionals: *Global strategies for environmental issues*, 12-15 June 1994. New Orleans, Louisiana. 266-276.

Welford, R. (1994). Barriers to the improvement of environmental performance: the case of the SME sector. In Welford, R. (ed). *Cases in Environmental Management and Business Strategy*. London. Pitman Publishing. 152-165.

Welford, R. & Gouldson, A. (1993). *Environmental Management and Business Strategy*. Pitman Publishing, London.

Whatmore, S. & Voucher, S. (1993). Bargaining with nature: the discourse and practice of 'environmental planning gain'. In Burgess, J. (ed). *People, economies and nature conservation*. Ecology and Conservation Unit Discussion Paper No 60. University College London. 91-111.

World Commission on Environment and Development. (1987). *Our Common Future*. Oxford University Press, Oxford.

3 The Involvement of Developers in Local Agenda 21

ALISON DARLOW AND NORMA CARTER
De Montfort University, Leicester, UK

INTRODUCTION

Over the past few years, sustainable development has become an area of increasing concern for local authority plans and policies. Despite this, much of the debate surrounding sustainability has concentrated on identifying a useful definition of the term. However, it is generally agreed that at the heart of sustainable development lie four principles: futurity, environment, equity and participation (Elkin, McLaren and Hillman, 1991). Perhaps a more useful definition of sustainable development is that it is about *'improving quality of life whilst protecting the environment for the current and future generations'* (Charter and Newby, 1995).

The debate surrounding sustainable development has intensified in recent years through the development of Agenda 21. Agenda 21, which was signed by 178 nations at the Earth Summit in Rio in 1992, is an agreement to work towards sustainable development. This commits each national government to report annually to the United Nations Association Commission on Sustainable Development on progress towards sustainable development. Chapter 28 of Agenda 21 refers specifically to local authorities. This outlines an important role for local government and calls for them to consult with their local communities, to produce a blueprint for the future of their area, widely known as a 'Local Agenda 21'.

Local Agenda 21 emphasises that under-represented groups should be able to have an input into the consultation process. These include groups such as women, children and youth, voluntary groups and non-governmental organisations, workers and trade unions, business and industry, cultural and ethnic communities, senior citizens, the unemployed and economically deprived and people with disabilities.

In 1994, the UK Government produced 'Sustainable Development: The UK Strategy' in response to Agenda 21 (DoE, 1994), which outlines how Britain proposes to make progress towards this goal. As yet there has been no formal

The Environmental Impact of Land and Property Management, Edited by Yvonne Rydin
© 1996, The Royal Institution of Chartered Surveyors

guidance from central government as to how local authorities should respond to Local Agenda 21.

The Local Agenda 21 Steering Group manages the initiative in the UK, and the Local Government Management Board co-ordinates responses from local authorities on behalf of the Steering Group. Many local authorities have made considerable progress with their Local Agenda 21 initiatives. Recent research has shown that 72% of authorities in the UK were committed to participating in a Local Agenda 21 process (Tuxworth and Carpenter, 1995).

The very nature of the development industry inevitably means that it will have a profound effect on the environment. In terms of the volume, location and type of development, this industry can and does have a large impact on the environment. However, despite the apparent incompatibility of the development industry with the aims of sustainable development, there does seem to be some attempt to 'green' the industry (for example, CIRIA, 1995). In addition, research has concentrated on the environmental awareness of the development industry (Jelley, 1993), and on how the industry can improve its environmental credentials (Rydin, 1994).

Therefore a policy for more sustainable development by the development industry may put more emphasis on the location, type and form of development, the re-use of existing buildings and the recycling of materials. Whether the impetus comes from central government regulation, through voluntary codes of practice, or is market-led, fundamental changes are clearly required within the industry if there is to be any hope of achieving more sustainable development.

Clearly the development industry will continue to have a major impact on the built environment, and particularly on the development of towns and cities. It will be especially significant in the production of more sustainable forms of development. Because of this important role, and also as the development industry forms a major part of the general business community, it seems important that the development industry is involved in the Local Agenda 21 process.

Bearing this in mind, the overall aim of this research project was to provide a better understanding of how the development industry (including development companies, their umbrella organisations and representatives) contributes to the Local Agenda 21 process. Recognising the importance of the current sustainability debate, and in particular relating to the Local Agenda 21 process, the project aimed:

1) To establish whether and how developers are involved in the Local Agenda 21 process.

2) To identify how practice could be improved from both the point of view of developers and local authorities, with regard to the Local Agenda 21 process.

3) To make recommendations as to how practice could be improved in the future to ensure appropriate dialogue between developers and local authorities in the Local Agenda 21 process.

The research comprised a national survey of developers and house builders, followed by a number of more in-depth case studies. A self-completion questionnaire was mailed to 552 developers and house builders, covering all the main property market sectors. A response rate of 36% was obtained, which is good for a postal survey of this type. Three case study areas were selected for further investigation. These allowed issues identified in the questionnaire phase to be investigated in more depth, and also to examine in more detail the opportunities available for developers in get involved in Local Agenda 21s.

SURVEY OF DEVELOPERS

The questionnaire obtained responses covering the entire breadth of the development industry. More than four out of five of the developers operated in the housing sector: 48% of respondents were solely house builders whilst 19% operated in areas other than house building. Around a third operated in both house building and other development sectors.

In terms of scale of operation, over half the sample considered themselves to be a regional organisation and just over a quarter claimed to operate on a national scale.

GREENING THE DEVELOPMENT INDUSTRY

In order to establish levels of environmental concern and action within the development industry, respondents were asked whether (to the best of their knowledge) their organisation had a written environment policy. In response, 17% stated that their organisation did have a written environmental policy, indicating a certain degree of greening within the industry.

Respondents were also asked if they were involved in any other forms of environmental planning (Figure 1). Involvement in development planning was indicated by the majority of respondents, and around a quarter were involved in Urban Development Initiatives, including City Challenge. Nevertheless over a quarter were not involved in any type of planning activity at all.

DEVELOPERS' AWARENESS OF LOCAL AGENDA 21

Awareness of Local Agenda 21 was relatively low: only 13% of developers that answered the questionnaire stated that they had heard of Local Agenda 21. This may not, however, give a true representation of the extent of awareness, due to the problems associated with the self-selecting nature of postal questionnaires. In addition, questionnaires were answered by a whole range of professionals within the development companies, ranging from Managing Directors to planning assistants. Therefore awareness of Local Agenda 21 is likely to vary widely across organisations.

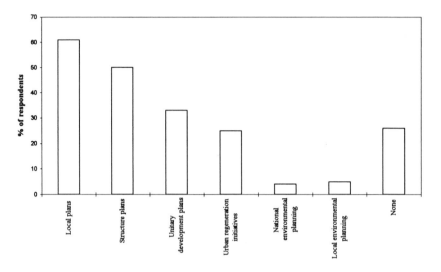

Figure 1 Involvement in environmental planning

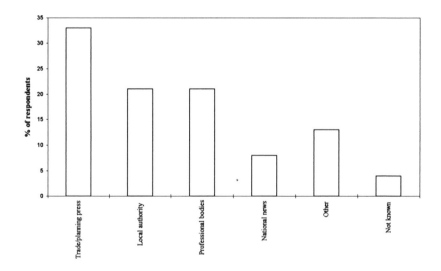

Figure 2 How developers became aware of Local Agenda 21

The main sources of information about Local Agenda 21 for the development industry were the professional press, especially the planning and building press. Other important sources of information included local authorities and professional bodies, with The Royal Institution of Chartered Surveyors and House Builders Federation mentioned in particular.

Of those developers that were aware of Local Agenda 21, only four were aware of any other organisation representing their interests within the Local Agenda 21 process. Organisations that were mentioned included the House Builders Federation, local Groundwork Trusts and the British Property Federation.

Certain types of developer appeared to be more likely to be aware of Local Agenda 21. For example, developers that were aware of Local Agenda 21 were more likely to be involved in house building whilst unaware developers had a heavier involvement in other market sectors.

Disparities were also evident when the two groups were compared in terms of involvement in other environment initiatives. Nearly a quarter (24%) of developers that were aware of Local Agenda 21 had a written environmental policy, compared with 15% of unaware developers. In addition, developers that were aware of Local Agenda 21 were more likely to be involved with other types of environmental planning. This perhaps indicates that a group of developers tends to be more active and more environmentally aware. Of course, it is likely that being active and aware will be easier for regional or national developers who would be able to dedicate staff to engage in these types of activities.

In addition, the large number of house builders who tend to be more aware, may have become so as a consequence of being involved in land-use planning. Due to constraints on their development activity through housing allocations in development plans, and their involvement since the early 1980s in development plan consultations, house builders have a particular history of experience and involvement.

DEVELOPER INVOLVEMENT IN LOCAL AGENDA 21

In terms of actual involvement in Local Agenda 21, only four developers had been asked to participate in a Local Agenda 21, and all had agreed to do so. This is equivalent to 15% of developers that were aware of Local Agenda 21, but only 2% of the total sample.

Of the developers that were involved in Local Agenda 21s, three considered that they operated on a national level, and one on the regional scale. Three of the developers were involved in a range of types of development, including residential, office, industrial and retail. One developer concentrated only on the housing market. There does not seem to be any clear link between volume of output and involvement. Although only one developer had a written environmental policy, all the developers involved in the process felt it was important for developers to be involved in Local Agenda 21. These developers clearly consider that being involved in Local Agenda 21 is both beneficial and of relevance to them.

In terms of how developers were involved, three respondents were involved in the process through round tables or working groups, with one respondent being involved though a local society.

Developers that had not been asked to participate were relatively mixed about whether they would do so if asked. Over a half (53%) of developers were not sure whether or not they would agree to participate, over a quarter (26%) would participate and 21% would decline the invitation. This may be due to an unwillingness to commit staff time and resources to something that does not appear to show an immediate benefit.

IMPROVING OPPORTUNITIES FOR INVOLVEMENT

Of the developers that responded to the survey, 45% thought that opportunities for developers to be involved in Local Agenda 21 could be improved. Suggested methods for improvement are outlined in Figure 3. These included involving developers more at the local level and actively seeking developer comments, more involvement through professional bodies, and more available targeted information.

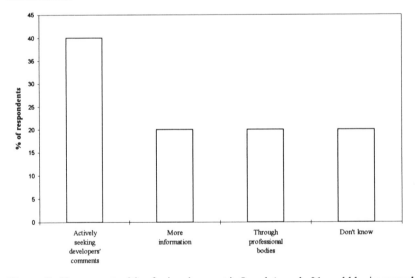

Figure 3   How opportunities for involvement in Local Agenda 21 could be improved

Developers were also asked if they thought that it was important for them to be involved in the development of Local Agenda 21s. Despite the low levels of commitment and involvement, 87% thought that it was important for developers to be involved. This could indicate, as the figures may suggest, that there is a high level of potential interest and concern in the development industry. However, this could also indicate that developers do not want to feel left out of any process that they could potentially be a part of, but would be unwilling to commit financial resources or staff time.

SUMMARY

The survey indicates that developers have some awareness of Local Agenda 21, gained mostly by reading the professional press and through contact with local authorities. However, awareness is on the whole low, and there is even less involvement in the initiative.

Developers that were aware of Local Agenda 21 tended to be larger, were more likely to be house builders, and also to be more active and environmentally aware than other developers. Of those developers that were aware of Local Agenda 21, nearly half felt it was important for developers to be involved in the process. Further questioning revealed that developers felt that it was important for them to be involved as they could inject some commercial realism into what could otherwise be seen as an 'airy fairy' exercise.

Upon further probing, developers that were aware of Local Agenda 21 recognised that they did have an important role to play in achieving sustainable development. Different perspectives were put on this. On the one hand, one developer argued that they would inevitably have a greater role in sustainability due to new planning legislation, as in order to get planning permission they have to demonstrate that a scheme is sustainable. On the other hand personnel within development organisations expressed the view that developers had been given much too free a rein so far, and that the development industry had been responsible for much unsustainable development and erosion of the environment. It was also perceived that only the right political will would change the development industry in any significant way.

The developers that have been involved in the Local Agenda 21 process, had all been invited to participate by the appropriate local authority. They tended to be larger developers, with regional and national impacts, and perhaps significantly all were active in the residential sector.

In terms of improving and increasing participation by developers, three ways forward were identified in the survey and also reinforced by further discussions with developers.

Firstly, the co-ordinators of Local Agenda 21 could be more proactive by actively seeking input by developers. This could take the form of co-ordinating an authority-wide forum of all those involved in construction and development, through to invitations to contribute directly.

Secondly, a need was perceived for more targeted information and advertising. Finally, greater input and co-ordination through regional and local branches of professional bodies were also called for.

THE CASE STUDIES

The case studies aimed to explore issues about developers and Local Agenda 21 in more detail, and in particular to look in depth at issues raised during the questionnaire phase of the research. Three case studies were selected (Leicester, Leicestershire and Leeds) on the basis of the level of progress with Local Agenda 21, geographical range, type of authority, a mixture of urban and rural areas, and the availability of information.

Leicester

Leicester's Local Agenda 21 has been renamed *'Blueprint for Leicester'*. The title is supported by a logo and strapline (*'Make your mark on Leicester'*). *Blueprint* is jointly co-ordinated by three organisations: Environ (a local environmental charity), Leicester City Council, and Leicester Promotions Ltd (an arm's length City Council supported organisation responsible for the promotion of Leicester).

The Environment City initiative has provided the starting point for Leicester's Local Agenda 21. Leicester was designated as Britain's first Environment City in 1990 (Wood, 1994). Leicester is generally regarded as fairly well advanced in the Local Agenda 21 process, and the city is acting as one of the four 'shadow' authorities in the LGMB Indicators Project. The indicators will play a key role in the monitoring of *Blueprint* (Environ and Leicester City Council, 1995).

Two key areas of opportunity have arisen for developers to input into this process. Firstly, through the Environment City Specialist Working Groups (SWGs) and secondly through a series of Target Groups.

The SWGs were intended to act as a microcosm of the Environment City initiative, encouraging partnership between different sectors. However, involvement from the business sector has been low. The SWGs that are most relevant for developers are those for the Built Environment and for Economy and Work. In terms of the Built Environment SWG, developers are represented by the House Builders Federation (HBF) and the Building Employers Federation (BEF). Representation on the Economy and Work SWG was limited to general business representatives such as the Leicestershire Training and Enterprise Council.

Developers were also able to contribute to the process through Blueprint Target Groups. Various initiatives have been implemented to improve input from the groups as identified by Local Agenda 21, including special events for targeted groups, umbrella business groups, mail-outs to individual firms. Leicestershire Economic Development Partnership has co-ordinated business responses for both the Leicester and Leicestershire Local Agenda 21s. However, there has been no actual input by individual developers or by developer organisations.

Leicestershire

Leicestershire's Local Agenda 21 has been called '*A Better Leicestershire*'. Although Leicester had already launched a Local Agenda 21 process, it was seen that there was also a need to produce a strategy for the whole of Leicestershire. All of the district councils in Leicestershire have participated in some way in the Leicestershire Local Agenda 21.

Leicestershire co-ordinates responses from the other district authorities through a county wide Local Agenda 21 forum, known as Forum for a Better Leicestershire (FABLE). A number of opportunities have arisen for input by developers into the Leicestershire Local Agenda 21 (see below). These are:

1) *Forum for a Better Leicestershire (FABLE)*
   Forum membership covers a wide range of groups, including businesses, all having an equal footing to ensure wide ownership. The regional branch of the HBF has been involved, although no developers have had direct involvement. Representatives are also able to have an input into the FABLE steering group.

2) *FABLE Action Groups*
   Membership of the Action Groups was determined by drawing from the membership of FABLE, as well as invitations to under-represented groups and experts. Some of the issues put forward from the Action Groups were particularly relevant for developers, such as the amount of built-up land in Leicester in 30 years' time.

3) *Business and Industry Target Group*
   Membership of the Business and Industry Target Groups was determined by Leicestershire County Council and Environ (on behalf of FABLE), identifying both local umbrella organisations and large local employers.

On the whole, no input from individual developers is evident, however the Regional Planning Officer from the House Builders Federation has been involved in FABLE and on the FABLE Building, Planning and Transportation Action Group.

Leeds

Whilst Leicester and Leeds are both Environment Cities, Leeds has developed a different approach to Local Agenda 21. As the whole Environment City initiative in Leeds is being used to work towards sustainable development, Local Agenda 21 and Environment City are very much seen as being the same initiative. Therefore, wherever possible, public consultation and participation are being carried out through existing fora and networks. This is distinct from the approach taken in Leicester, where the Local Agenda 21 process is seen, and being promoted, as a separate project that is needed to supplement the participation and planning work already taking place through the Environment City initiative.

Whilst the Leeds Environment City initiative is co-ordinated by a small secretariat, it is essentially a partnership between three bodies: Leeds City Council and its Green Strategy Implementation Team, the Leeds Environmental Business Forum and the Leeds Environmental Action Forum. In addition, Local Agenda 21 issues are being dealt with through Environment City Specialist Working Groups, in particular in liaising with other Environment Cities to produce a set of joint Sustainable Development Indicators.

There are two main opportunities for developers to become involved in Local Agenda 21 in Leeds. Firstly, the Environment City Specialist Working Groups, which are similar in structure and purpose to the Leicester SWGs, provide a means of participating in the Local Agenda 21 process. No individual developers or developer organisations are involved, however, although the Director of the Leeds Development Agency is a member of the Economy and Work SWG.

The second means of participating is through the Leeds Environmental Business Forum, which is regarded as spearheading business involvement in the Local Agenda 21 process. However, the Forum does not at present have any developers as members, although the development industry is regarded as represented to some extent by civil engineers, a planner, legal firms who deal with property as well as a number of architects. The forum is open to new members, but developers have not yet been targeted as a member group.

COMPARISON OF APPROACHES

A range of opportunities does exist for input by the development industry into the case study Local Agenda 21 processes. No individual developers have actually been involved, although some of these opportunities have been utilised by regional branches of national developer organisations. All the Local Agenda 21s identified business as a key area to target, and therefore developers are represented to some degree by general business associations. For example, in Leeds businesses have been given special prominence through the establishment of the Leeds Environment Business Forum.

In Leicester and Leicestershire, new public consultation procedures have been adopted for Local Agenda 21, and a range of opportunities could be identified. Despite this no individual developers were being involved, although the regional branches of the HBF and the BEF had some involvement. However, problems were highlighted by the HBF due to the amount of time that needed to be devoted to these initiatives, and the large number of potential Local Agenda 21s to cover. It should be noted, however, that the main part of the consultation exercise of Leicestershire's Local Agenda 21 is still to be completed.

Leeds' approach to Local Agenda 21 is very different from the approaches of Leicester and Leicestershire, with existing mechanisms and fora providing the basis for the initiative. However, opportunities did exist for developers to have an input into the process, predominantly through the Environment City SWGs and also through the well-established LEBF. Nevertheless, there still seems to have been little participation by developers. In addition, it seems that there has been no input by developer organisations, although business as a whole is represented by other general business organisations.

In terms of improving participation levels, the co-ordinators of *Blueprint for Leicester* argue that that locally based developers should be given the opportunity to be involved in the process, as should any other locally based organisation. Therefore publicity should be aimed in particular at all businesses, suggesting ways of being involved. This should be a fairly flexible approach outlining a number of different ways to take part.

In addition, the co-ordinators believe that developers should be involved early on due to their large environmental impact and influence on the future of cities in terms of the use of land, the production of housing and the protection of green space. Their views however, as with all businesses, should be balanced with other views from the wider community.

In terms of the Leicester Local Agenda 21 process, it was felt that improving levels of input from developers could potentially be achieved by establishing higher level links with organisations. Often approaches have been made to firms by medium ranking officers, and a more senior level interface between the developer and the co-ordinators of Local Agenda 21 could make a difference to the level of actual participation.

With regard to the use of umbrella organisations and professional bodies to co-ordinate input from developers, this was seen as not being particularly useful because they may only represent the views of large firms and only on a broad regional basis. The regional arm of the HBF has been involved to some extent, although time constraints and other commitments have severely limited this involvement.

DISCUSSION OF FINDINGS

On the whole, despite levels of involvement by developers or developer organisations being low, all the case study Local Agenda 21s had specialist areas of input for businesses, as well as more general means of participating.

The low participation rates for developers in the case studies indicates a number of possible factors:

- Lack of publicity or information being given to the appropriate bodies.

- Lack of direct encouragement to participate from the Local Agenda 21 co-ordinators, or from umbrella organisations representing business input.

- Reluctance on the part of the developers to become involved, possibly due to a lack of available time, and a higher priority being given to profit based objectives.

Certainly it seemed to be the case that the co-ordinators of Local Agenda 21 felt that local developers should be involved in the process. Whether this input should just be as a part of the wider business community was not always clear. A question arises as to whether the particular significance of the development industry in the production of the built environment warrants a special place in the Local Agenda 21 process.

In terms of improving opportunities for developers, it was felt that there was only so much the co-ordinators could do. If the opportunities to participate were well advertised and clear encouragement was given, then developers had to take the initiative and recognise the importance of being involved for themselves. As one developer stated, lack of participation now will only lead to problems further down the line, as developers get more and more out of touch with the sustainability debate and other local environmental issues.

CONCLUSIONS

The issue of sustainable development in general and Agenda 21 in particular is receiving a steadily increasing amount of attention internationally, nationally and locally. Local Agenda 21 processes are now being developed by an increasing number of local authorities.

Local Agenda 21 seeks to work towards sustainable development. The importance of consulting with local communities is highlighted and business has been targeted as one of a number of priority groups. Considering the role of developers as part of the wider business community, and also as a major influence on the type, scale and nature of future development, it seems important that they should have some input into Local Agenda 21.

Levels of awareness and participation in the industry are, however, low. Just over 1 in 10 of developers surveyed were aware of the initiative, with only 2% being involved. Despite this apparent lack of involvement, developers that were involved generally thought that it was important for developers to be included. Ways of increasing involvement were suggested, and a strong onus was put on the co-ordinators of Local Agenda 21, both by inviting responses and input directly from developers and by improving the quality and quantity of information and advertising in communication channels used by developers.

In terms of the case study Local Agenda 21s, opportunities did exist for developers to become involved in both the development of the Local Agenda 21 process and in the community participation phase. Nevertheless, very few of these opportunities were taken by developers. More general representation of the business community as a whole was evident through organisations such as the CBI and the local Chambers of Commerce. It would seem unlikely, however, that these organisations would be able to represent a specific developer point of view.

A number of issues may be identified as arising from the findings of the research project.

Firstly, developers clearly have major influence on the type, scale and nature of future development and will play a major role in achieving more sustainable development. It is generally agreed that they should have an input in Local Agenda 21. It is unclear, however, whether because of their particular significance in the production of the built environment they should still be treated as part of the general business community.

Secondly, it is apparent that both developers and the co-ordinators of the Local Agenda 21 see the other side as not being proactive enough. Developers tend to want to be spoon-fed, whilst co-ordinators think they already provide opportunities for developers to become involved. Perhaps the benefits of participation by developers are not felt significant enough by either side to warrant the necessary staff input. Lack of input by developers could equally be seen by either side as 'no great loss'. On the one hand, the co-ordinators of Local Agenda 21, who are more likely to be environmentalists at heart, may see that no input from developers could result in a greener Local Agenda 21 strategy. On the other hand, developers are likely to perceive Local Agenda 21 as not being important or relevant to them.

Thirdly, there is an important issue regarding the appropriate level of involvement for developers. Should national developers be involved at the local level? If so with which Local Agenda 21s should they be involved: where the developer is based or where they are currently operating?

Finally, umbrella and professional organisations could be a key to better input. Nevertheless, reservations should be expressed about whether the regional tier of such bodies can usefully play in a role in local initiatives, and also whether they have at the moment the right attitudes, co-ordination skills or even the necessary staff time.

A debate emerges as to whether developers should be encouraged to be involved in Local Agenda 21. A Local Agenda 21 that does involve developers may not necessarily be as 'green' as a strategy as one that has not involved them, but it may have a better chance of being translated into policy and practice. In taking developers' comments on board, we are not necessarily talking about compromise as much as greater consultation. As one developer in support of greater participation stated, lack of participation from developers now will only lead to problems further down the line, as developers get more out of touch with the sustainability debate and local environmental issues. Rather than sweeping areas of potential conflict under the carpet they should be focused upon. Frank debate and discussion are needed to identify the best ways forward.

Awareness raising and direct encouragement to participate are clearly keys to greater involvement from the business community as a whole. It is imperative that local business organisations become involved in Local Agenda 21, because of their important contribution to the local economy and potentially to local sustainability.

Local Agenda 21 will not be implemented as fully as possible unless the business sector is involved alongside other interests and the community at large. Greater involvement of developers, particularly those operating at the local level, could better reflect the full range of views across all sectors of the community, as well as providing added value in terms of raising the environmental awareness of developers.

The importance of the involvement of the business community in any drive for local sustainability has long been recognised. Actually achieving the rates of participation may be an altogether different story. As far as developers are concerned at least, it is likely that this industry will continue to affect our ability to become more sustainable. We wouldn't be so naive as to suggest greater involvement in Local Agenda 21 will be a recipe for instant success: changes in legislation and central government ethos, as well as a greater degree of local autonomy, are all needed. However, in the mean time sticking our heads in the sand and hoping these difficult issues will go away will simply not work.

ACKNOWLEDGEMENTS

Thanks are due to many individuals in a wide range of organisations and companies who generously gave their time and provided information. Special thanks go to the developers and house builders who responded to the questionnaire and to those individuals who generously gave advice and comment at various stages of the work. In particular, the House Builders Federation and the National House Building Council assisted at the sampling stage and the co-ordinators of Local Agenda 21 provided documents and gave time for interviews.

Finally thanks must go to The Royal Institution of Chartered Surveyors who funded the project through their Environmental Research Programme.

REFERENCES

Charter, S. and Newby, L. (1995). *The Evolution of a New Approach to Local Environmental Challenges with Special Reference to Local Agenda 21 and the Landuse Planning Function.* Paper presented to Environmental Research Conference. Lodz, Poland. July.

CIRIA (1995). A clients guide to greener construction, *CIRIA Special Publication 120.* CIRIA, London.

Department of the Environment (1994). *Sustainable Development: The UK Strategy.* HMSO, London.

Elkin, T., McLaren, D. & Hillman, M. (1991). *Reviving the City: Towards Sustainable Urban Development.* Friends of the Earth and the Policy Studies Institute, London.

Environ and Leicester City Council (1995). *Indicators of Sustainable Development in Leicester: Progress and Trends*. Environ and Leicester City Council, Leicester.

Environ, Leicester City Council, Leicester Promotions (1995). *Blueprint for Leicester: Findings Report*. Environ, Leicester City Council & Leicester Promotions, Leicester.

Estates Gazette (1994). Sustainable Development, *Estates Gazette*. **9545** 145-146.

Forum for a Better Leicestershire and Environ (1995). *Quality of Life and Environmental Issues in Leicestershire: A Survey of Local Attitudes for Leicestershire's Local Agenda 21*. FABLE, Leicester.

Forum for a Better Leicestershire and Environ (1995). *The Leicestershire Local Agenda 21 Communications Strategy*, FABLE. Leicester.

Jelley, D. (1993). Implications for Commercial Property Development of the Environmental Debate. Unpublished Post-graduate Dissertation, De Montfort University.

Royal Institution of Chartered Surveyors (1994). *Financial Incentives for Greener Homes*. RICS, London.

Rydin, Y. (1994). The Greening of the Housing Market. In Bhatti, M. et al. (eds) *Housing and the Environment: A New Agenda*. The Chartered Institute of Housing, London. 128-140

Tuxworth, B. and Carpenter, C. (1995). *Local Agenda 21 Survey*. University of Westminster, London.

Wood, C. (1994). *Painting by Numbers*. RSNC, Lincoln.

World Commission on Environment and Development (1987). *Our Common Future*. Oxford University Press, Oxford.

World Conservation Union, United Nations Environment Programme and the World Wide Fund for Nature (1981) *Caring for the Earth: A Strategy for Sustainable Living*. Gland, Switzerland.

4 An Investigation into the Application of Quantity Surveying Skills in the Cost Management of Environmental Issues

C. L. PASQUIRE
Loughborough University of Technology, Leicestershire, UK

INTRODUCTION

The research was conceived from a desire to show that quantity surveyors could and should have a role to play in the environmental issues that confront our future and the construction industry in particular. This gave rise to a dual vision for the work; firstly, that quantity surveyors could contribute using their unique cost analysis and management skills and that they should contribute for both moral and commercial reasons.

It came as no surprise to find that the construction industry was again lagging behind others in their approach and understanding of environmental issues. This was largely due to few of the parties being willing or able to provide services for which they were not being paid. This does not mean to say that these services were or never will be needed, indeed this research showed that the whole green issue was growing in momentum and that a market was forming. It is speculated that this market will grow rapidly once established and that opportunities missed now will be hard to claw back in the future.

It could not be said, however, that the UK construction industry does not acknowledge the increasing public awareness of environmental issues and it is taking measures to help reduce the impact of this awareness. In addition to responding to public demand, the industry has a duty to educate the public and rationally inform the environmental debate. In some quarters a considerable environmental effort is being made but all too often environmental friendliness is blatantly used as a marketing tool for construction products and companies.

The current green bandwagon has gained impetus in the construction industry from several directions :

The Environmental Impact of Land and Property Management, Edited by Yvonne Rydin

- Direct competition from Europe
- Growing commercial concern over environmental risk and liability
- Current and forecasted consumer sophistication and demand
- Anticipated legislation, standards and taxation introduced from Westminster, Brussels and international assemblies.

There is a distinct lack of adequate and rational environmental information. Consequently there is a clear role for the industry to undertake its own initiatives, to make available sound information and incentives for change. The RICS and various divisions have begun to address a variety of environmental issues in the professional press. The scarcity of contributions from quantity surveyors on the rapidly growing topic of the environment would suggest the quantity surveyor has little interest and no perception of what is involved. It was the intention of this research to demonstrate that quantity surveyors have an opportunity to turn existing skills and knowledge into services for the customers of tomorrow's profession by establishing a growing niche market in the environment. As Baker (1993) stated 'as soon as choice and cost become variables in the same property related question, the quantity surveyor has a contribution to make'.

The basic research methods used were a comprehensive literature search along with interviews held with a variety of environmental consultants, clients and quantity surveyors.

This paper can only briefly outline the findings of a considerable piece of research which is itself the subject of a much longer and more detailed document in the process of being prepared at Loughborough University (Pasquire and Plunkett, 1995).

ENVIRONMENTAL KNOWLEDGE REQUIRED

In order for the quantity surveyor (or any other professional) to provide services to the construction industry in the environmental field, they must possess a specific and detailed knowledge of the following matters.

Environmental issues in construction

What are the environmental issues that face the construction industry? These were identified and categorised by CIRIA (1993) as:

Energy use, global warming and climatic change: This revolves around the 'greenhouse' effect and the need to reduce the emission of greenhouse gases into the atmosphere. In construction, these arise from the generation of energy for use in the manufacture of products and materials, during construction and during the use of the occupied building. They are also released during the chemical processes used for the manufacture of materials and products. The natural balance of the

Earth's atmosphere has also been disrupted by deforestation, the construction industry being a substantial consumer of timber and paper products.

Resources, waste and recycling: This issue can be considered as a resource cycle. Vast quantities of natural resources are used in construction of buildings, roads and infrastructure, but their eventual demolition or redevelopment can produce waste which can be recycled on other projects. Unfortunately, the fragmented nature of construction development means that the practical extent of recycling is constrained. The issue of resource sustainability is of utmost importance; for example, the use of tropical timber, a potentially sustainable resource, is exploited in some parts of the world in an unreplenishable way. Recycling can be a cost effective option and according to CIRIA (1993) there is considerable potential within the construction sector for recycling and the reduction of over-consumption of construction resources by a variety of means.

Pollution and hazardous substances: Broadly speaking, pollution from the construction industry can be considered in four categories:
- atmospheric pollution caused by dust from demolition and the movements of vehicles, burning of waste releasing toxic gases and solvent releases from paint, strippers, degreasers etc.   *Extend with UPS 3.7*
- water pollution caused by spillage or silt run-off as a result of earthworks.
- noise and vibration pollution caused by the construction process itself and in particular piling, demolition, ground consolidation, blasting and the use of heavy plant and equipment. This type of pollution can lead to damage to buildings and sensitive equipment, e.g. computers.
- construction materials may themselves be hazardous brought about on installation, when worked, when they decay and when disposed of. The largest area of concern is currently land contaminated by hazardous materials left behind from industrial developments. Land may also be contaminated by naturally occurring geological phenomena and historic mining activity.

Internal environment: This includes such issues as Sick Building Syndrome (SBS), Legionella and radon gas, all of which place restraints upon the design and construction of buildings.

Planning, land use and conservation: A wide range of environmental issues exist in connection with the interaction of the land use planning system and the construction industry. The two principal issues are improving the management of the building process and enhancing communications between the general public and the parties responsible for the development. Construction activity and subsequent building/land use impact heavily upon transport movements. Specific areas of environmental concern cover the impact of mineral extraction and development projects, infrastructure issues concerning the location of waste disposal/treatment facilities, dereliction and contamination, and the designation of areas for nature conservation and specialised habitats. See also CIRIA (1987); DTI (1994); DOE (1990); and Bright (1991).

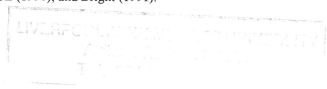

It can be seen that there is a vast array of environmental issues that confront the construction industry. It is clear that businesses will have to become more aware of these issues in local, national and indeed global terms. In order to respond and manage these issues, a variety of skills are and will be provided in the form of environmental services to deal with and minimise the problem. It must be recognised, too, that although the issues can be identified and classified separately, in reality they are all interdependent and can not be considered in isolation.

Environmental legislation affecting the UK construction industry

The legislation concerning environmental issues is broad and varied and there is no single statute governing the environment as a whole. It must not be forgotten that more legislation is emanating from Europe and that ultimately there will be global laws and conventions. In the UK, a legal definition of 'environment' has been given in the Environmental Protection Act 1990 as 'The environment consists of all, or any, of the following media, namely, air, water and land; and the medium of air includes the air within buildings and the air within other natural or man-made structures above or below ground'.

An outline of current UK legislation

The main UK Acts of Parliament were listed by CIRIA (1994) as:
- Alkali Works etc. Regulation Act 1906
- Building Act 1984, Building (Scotland) Act 1970 & Building Regulation 1994
- Clean Air Act 1993
- Control of Pollution Act 1974
- Environmental Protection Act 1990
- Health and Safety at Work etc. Act 1974
- Occupier's Liability Acts 1957 & 1984
- Planning (Hazardous Substances) Act 1990
- Town and Country Planning Act 1990; (Scotland) 1972, Planning and Compensation Act 1991, Planning (Consequential Provisions) Act 1990
- Water Resources Act 1991
- Water Industry Act 1991

Of these the most significant were:

Environmental Protection Act 1990: This Act governs a wide range of potential polluting processes; controls the disposal of waste on land and has modernised the law of 'statutory nuisances' creating a duty on local authorities to register contaminated land.

Water Resources Act 1991: This lays out the statutory system for the management of pollution of inland and coastal water by the National Rivers

Authority. It also controls discharges to surface waters by a system of consents. Discharge of trade effluent to sewers, however, is regulated under the *Water Industry Act 1991*, also by consents.

Health and Safety at Work Act 1974: This imposes a duty on the employer to provide a safe working environment. See also *Control of Substances Hazardous to Health Regulations 1988* (COSHH) which aims to protect against health risk from exposure to hazardous substances at work, and *Construction (Design and Management) Regulations 1995* (CDM) detailing responsibility for health and safety in the construction industry.

The Town and Country Planning (Assessment of Environmental Effects) Regulations 1988: Although not fundamentally an environmental statute, this introduced the initiative of Environmental Impact Assessments. It allows local authorities to exercise indirect controls over development by establishing environmental factors as 'material' planning considerations. How the actual construction is carried out is the subject of the *Building Act 1984* supplemented by the Building Regulations and Approved Documents. Although limited in the environmental field, these have tended to concentrate on energy conservation and the use of environmentally friendly building materials.

An outline of European policies

EU legislation is enacted mainly by directives or regulations. Directives have no effect until the member state has passed the relevant implementing legislation. Regulations, on the other hand, are laws in their own right and do not require implementing legislation. They are effective immediately. On the whole, most environmental legislation is considered by directive, e.g. landfill of waste, tax on carbon dioxide emissions, energy efficiency and the promotion of renewable energy and the green paper Remedying Environmental Damage 1993. Regulations have been used very sparingly with an exception in the phasing out of the use of CFCs. There is no institution which enforces EU environmental legislation. The only remedy is to take the government of the member state to the European Court, if directives are not implemented within an appropriate time scale or in a proper manner.

BS7750 Environmental Management: Although not legislative, this was developed by BSi as a shadow to the EU draft Eco-audit Regulations. It aims to protect the environment through a quality management approach and can be interpreted as a sort of 'green badge' of acceptability for companies awarded the standard. The purpose of adopting an Environmental Management System (EMS) can include cost reduction, image enhancement, customer/client demand, investor's demand and insurance.

The Environmental Bill 1995: Creates an Environmental Protection Agency and brings together the NRA, the Water Resources Act and HM Inspectorate of Pollution. Introduces extensive provisions relating to contaminated land and coal mines and dealing with countryside policy, minerals legislation and waste management.

It can be seen that there is a fairly bewildering array of environmental legislation and regulation which affects the construction industry both directly and indirectly. The expertise that QSs have learnt through the interpretation and implementation of highly complex construction contracts must place them at an advantage when attempting to interpret and assess the effect of environmental legislation.

ENVIRONMENTAL SERVICES AND THEIR PROVISION TO THE CONSTRUCTION INDUSTRY

The identification of the range of environmental services available was the next step in the research. Finding out what was already being provided gave a basis for establishing whether a role exists for the quantity surveyor. These provisions were identified by interviewing self proclaimed Environmental Consultants as follows:

Corporate environmental strategy: The definition of environmental objectives of a company for its employees, clients, shareholders and the public in the format of a comprehensive environmental policy.

Environmental auditing: A systematic, documented, periodic check on the environmental performance of an organisation, the management and the equipment. It facilitates management control of environmental practices and evaluating compliance with policies or good practice guidance or legislation. See Wood (1993) Town & Country Planning Regulations 1988.

Environmental impact assessments: These were introduced into British law in 1988 as the result of an EU directive (RICS, 1993). It is a procedure whereby the likely effects on the wider environment of proposed developments are contemplated as part of the planning and design phases. Consideration may be extended through the construction phase and into the completed project phase, then represented by means of an environmental statement.

Water pollution: This is a highly complex area involving the assessment of the effects of developments on the aquatic environment in and around sites in a wide range of geological strata, during such operations as mineral extraction, landfill and the development of ground water supplies. In addition, the quality of the water maintained within the building should be monitored and consideration given to the disposal of aqueous waste, principally the characterisation of effluent and its treatment.

Waste management: Becoming increasingly important and revolving around waste minimisation. On-site waste management comprises source reduction, re-use, recycling treatment and disposal. Primary methods of waste disposal in the UK are currently landfill, incineration and treatment.

Contaminated land: The investigation into contaminated land involves an examination of all matters relating to health and safety, pollution control, land drainage, compliance with legislation, environmental monitoring, geotechnical engineering, re-vegetation and re-development. Once the contamination problems are identified, alternative clean-up technologies can be explored and a solution designed. This exploration process will include assessments of effectiveness and cost. See Hawkins (1993) and Baker (1996).

Ecological and land management surveys and evaluations: Assess the impact of proposed developments and include land use surveys, landscape design, visual amenity, impact prediction, migration measures, advice on both restoration and cost effectiveness after use.

It would appear that the type of services currently being offered centre on a very wide range of highly specialised and technical skills, much of which are outside the existing, traditional quantity surveyor's abilities. It would seem unlikely that any existing professional has the skills necessary to perform all of the environmental services currently being offered. This means that there are two main thrusts to the opportunities afforded. Firstly, to manage the full spectrum of environmental services, buying in the necessary expertise as and when required, or to claim a niche market by becoming the sole supplier of a highly skilled service which underpins many of the current environmental services required. It is this latter that is the focus for the research undertaken for this paper.

ENVIRONMENTAL CONCERNS AND THE CLIENT'S ATTITUDE

In order to demonstrate an existing and future market, it is necessary to identify who requires environmental services within the construction industry and why.

Identification of clients

The parties involved with construction who could be identified as possible environmental services clients were identified and classified in the following sub-groups: Investors, Developers; Third Parties (solicitors, accountants, surveyors etc.), Local and Central Government and Others (water companies, construction companies etc.). This classification was derived to reflect the varying involvement in construction and attitude of the client sub-groups towards both construction and environmental issues (CSM, 1994).

Current approach to environmental issues

It seems that, although not a priority, environmental concerns were becoming increasingly apparent in client needs. It was found that, of the small sample of clients interviewed, 78% had an 'environmental policy' of some sort. The consensus seemed to be that wherever practicable, the companies incorporated environmental skills into their work by training existing employees or changing the company structure to include an 'environmental unit' or an environmental officer. Where the in-house expertise was insufficient environmental consultancies were approached for advice, e.g. building on contaminated land, carrying out site investigations, environmental impact assessments and audits etc.

Motivation behind the commissioning of environmental work

This was found to be twofold, firstly there was some tangible 'payback' either through an enhanced corporate image or real savings incurred, and secondly the motivation was a result of increased legislation. In the case of banks, legislation has resulted in a rise in borrower default due to environmental liabilities. It was disappointing to note that, where corporate image was a motivating force, the commitment was superficial with little if anything done to enforce policies implemented.

An examination of the motivation for commissioning environmental services provides the first clear indication that quantity surveyors do have a role to play in environmental issues. The motivating forces are cost and legislation, two areas where the quantity surveyor can claim a combined expertise above and beyond all other professions. However, existing analytical skills will need to be developed to envelop new law and technology.

QUANTITY SURVEYORS AND THE ENVIRONMENT

Although it can be seen that the construction industry is slow to embrace environmental matters, to assume these issues will not be of importance or provide a wide ranging market of opportunity is very short sighted indeed. I would speculate that within less than twenty years every aspect of the construction industry, and consequently the roles of those employed within it, will be significantly affected by not just environmental legislation but also by a changed public attitude towards the use of natural resources and the countryside. It is therefore essential that the quantity surveying profession makes the effort to get in front now if it wants to remain an influential profession for the future. But, how can this be done without a substantial investment of time and money? The primary source of expertise must be developed within current degree education, courses should be providing the foundation of a new direction for the profession. This can then be followed up through post-graduate training and Continuing Professional Development (CPD). Aside from this, the profession must look to the enhancement of its existing skills to gain a foothold in the market. Once momentum begins to build and fees can be earned, new skills can be developed within consultancies.

The cost management skills of a quantity surveyor are wide rang
the following services: development appraisal, cost analysis and
cost comparisons, cash flow forecasting and cost control.

All these are undertaken to a technical specification and within a legal
framework. The skills that underpin these services include: ability to measure and
quantify the project in a variety of descriptors, ability to interpret client
requirements and provide alternative solutions, appreciation of the difference
between cost and value, ability to track and organise data, and ability to anticipate
and project data into the future.

The knowledge that is needed to enable the execution of cost management
services includes: technology including performance, law, finance including
investment, taxation, grants etc., and how to find relevant information on these
subjects. These are all aspects of professional advice that are needed for
environmental services.

CURRENT ATTITUDE OF QUANTITY SURVEYORS TOWARDS ENVIRONMENTAL MATTERS

Only twenty quantity surveying or multi-disciplined firms offering any sort of
environmental service could be traced. Of these only seven were prepared to
contribute to this research. It can be seen therefore that the profession does not at
present involve itself in the environmental market to any great extent. The
pervading attitude of those interviewed showed that quantity surveyors had a
belief that they were 2 or 3 steps removed from the position of making influential
environmental decisions. These were generally thought to be made by the client or
designer before the quantity surveyor was involved. Most agreed that the quantity
surveyor could have a role but without a demand for the services it was difficult to
push the development of expertise forward.

Optimistically the view of the future was that demand would grow as a result of
increased legislation both within the UK and especially outside the western world
where big clean-up operations are anticipated. It appeared that firms involved
with the civil engineering and building refurbishment sectors of the industry were
much more actively involved in the provision of environmental advice than their
new build colleagues.

EXISTING DEMAND FOR ENVIRONMENTAL SERVICES FROM THE QS

There were five main environmental areas in which services were offered by the
firms interviewed. They were:

Contaminated land: Mainly involved at feasibility stage providing cost advice
on the cleaning up operations, suggesting and costing alternative solutions,
advising on site value before and after cleaning as part of development appraisal.

efficiency and global warming: The most common service was the
60 use of building services particularly in terms of energy use, emissions and
Legionnaire's disease. Systems were examined to ensure energy efficiency,
phasing out the use of CFCs and banning them in new development. The use of
halon in fire extinguishers had been addressed in one instance.

Waste management: This was seen as a potential growth area in terms of
recycling, reuse, energy from waste and renewable energy schemes. The two main
methods of disposal were landfill (involving excavation and lining of landfill site
and dealing with leachate) and incineration which required built structures to
house plant. One quantity surveyor firm felt they had indeed gained a competitive
edge from working with a major waste disposal company who had no experience
in building construction and refurbishment. Both parties had put a lot of effort
into learning about the other's business and the quantity surveyors concerned felt
they now had quite unique environmental expertise to offer to other clients.

Environmentally friendly and recyclable materials: It was recognised that the
quantity surveyor generally was not the specifier, however, there was thought to
be scope for advising, cost-in-use studies and costing of energy efficient materials.
The use of recycled and recyclable materials in refurbishment projects was a
growth area. Many materials installed could be pre-sold for recycling at the end of
their life span and a fairly large market is growing in recycling stripped out
materials. There is a growing requirement for the listing of hazardous materials
and this easily falls within the remit of the specification or contract documentation
preparation that quantity surveyors already undertake.

Timber from sustainable sources: Two of the quantity surveyors interviewed
had taken a strong stand on this and were actively influencing their clients to use
hardwoods from replenishable sources or fruit woods which are not only
sustainable but recycling a waste product of fruit growing (orchards are cleared
every thirty years).

Other: Radiological Assessment is considered a growth area for quantity
surveyors with Civil Engineering training. The construction and decommissioning
of nuclear power stations require careful considerations of the future use of such
structures and the ground upon which they are built.

The provision of environmental services by quantity surveyors is in its infancy,
however it might be useful at this stage to identify the clients commissioning
quantity surveyors to undertake environmental services. The government and its
agencies were identified as a major provider of work, followed by private and
public utilities and the National Rivers Authorities, the World Bank and waste
management companies. Pharmaceutical companies were also environmentally
aware as their products can have dangerous by-products . Other clients were large
plcs who wished to be seen, and presumably had a genuine desire, to be green.
These would include multi-national, blue-chip and major retail organisations.

THE DEVELOPMENT OF NEW AND EXISTING QS SKILLS

The existing quantity surveying skills identified above provide an excellent basis for the provision of auditing and analysis services in any construction related field. The main construction related environmental matters include:

- factors affecting the feasibility of land development including dealing with contaminated land and environmental impact assessment (EIA)
- factors which affect the building design including environmentally friendly materials and energy efficiency
- factors which affect the construction process including the control and reuse of construction material waste
- factors which affect the building use including protection of the environment from the processes and activities undertaken within the building - these would normally be identified during the EIA.

Many of these activities involve much the same technology and cost already familiar to the quantity surveyor - excavating for landfill is very little different than any other large scale excavation, building for incineration processes is very little different from building any other industrial structure. What we really have is a new terminology and an added layer of awareness not previously required. There is very little reason why quantity surveyors should not incorporate this environmental awareness into all their existing services - except that this would be construed as doing more for the same fee. But the construction industry is constantly changing, we all have to learn new skills and technologies all the time - remember SMM7? It involved a substantial degree of retraining in order for quantity surveyors to carry on providing an existing service. The body of knowledge within the quantity surveying profession is being added to all the time and quantity surveyors are expected to be up to date on, for example, new Building Regulations and new forms of contracts, providing an improved service for the same (or less) fee. By expanding existing knowledge of environmental issues the quantity surveyor can easily broaden existing services. The addition of this dimension would hopefully give a competitive edge when presenting bids for commissions. If the profession is to form a niche market in this way, the quantity surveyor *must* broaden their service base and develop new and existing skills.

The recent environmental movement provides new opportunities for the quantity surveyor. New skills are essential and need to be identified for future development of the quantity surveyor in the environmental field. There are three main areas which specify how the quantity surveyor could enhance their skills within the environment: risk analysis, valuation of amenities and facilities management including life cycle costing.

Risk analysis

'No risk' is not an option, as any option selected will involve certain environmental risks. A common misconception when considering environmental issues is that the most expensive option is the best choice. Exploring potential risks at the earliest possible opportunity leads to prevention rather than cure. An environmental screening exercise can categorise projects in accordance with their detrimental effects on the environment. This process establishes whether a full assessment or audit is necessary. An audit report may take the form of a matrix-based risk review. Costs can be assigned to the values in the matrix, calculated from the capital cost of construction, loss of revenue, penalties and the like. There are wide range of environmental issues which provide a choice, and hence where risk analysis can be applied. For example, environmental assessments and their mitigation measures, the Integrated Pollution Control provisions of the EPA (1990), waste and waste disposal options. In addition, contaminated land can be subjected to risk analysis, examining the cost implications of development options and consequently the selection of remedial options. Under BS7750, risk analysis can be related to decision making in the assessment of environmental effects. (See also Flanagan and Norman (1993), Perry and Thompson (1992), Pritchard (1994), Denner (1994) and Baker (1995).

Valuation of amenities

The problem of costing the value of amenities can be answered in the form of existence values (Hopkinson, 1990). The benefits of environmental improvements can alter slightly with increases or decreases in provisions and this can be identified by policy-makers in adopting a 'willing to pay' (WTP) strategy: consulting the general public on the qualitative factors that are generally ignored and on what they are willing to pay to prevent or promote a change in level of an environmental resource. Cost benefit studies using comparative values are possible options open to the QS in extending their traditional skills.

Facilities management and life cycle costing

Facilities management provides a one-stop management service to the client from conception to demolition/redevelopment (Brandon, 1990). If, in addition to this, a QS incorporates an environmental approach, a more marketable service would be available. Life cycle costing can address the area of energy efficiency in buildings, the energy used when the building is occupied and the energy used in the production of construction materials. In the latter case it involves the identification of the environmental effects of the materials and assesses the methods of extraction and the extent to which they pollute the environment.

None of the techniques described is new. Few of them have been adopted, however, by any party within the construction industry to any degree as they have been viewed, again, as an additional service for no extra fee, rather than methods of enhancing existing services. What is advocated here is the extension of existing techniques (although largely unfamiliar to QSs) into new areas. This requires a broadening of knowledge and an appreciation of a new set of influencing factors.

For example, in valuing an amenity, the loss of a family picnic area may be considered in terms of the costs for a family day out at a theme park in lieu.

THE FUTURE FOR QUANTITY SURVEYORS WITHIN THE ENVIRONMENTAL FIELD

In reflecting on the research so far, what has to be considered is, is there a future for the quantity surveyor within the field of environmental management? Furthermore, what measures are necessary to improve the quantity surveyor's position and opportunities in this field?

The profession foresees that the future demand for the quantity surveyor's skills in the environmental field will increase relative to the construction industry pulling itself out of recession. Although the traditional concept of a quantity surveyor is thought to be disappearing a more management based role could be adopted by existing quantity surveyors.

The quantity surveyor is not required to become an expert in environmental matters, but needs to have a basic understanding of the circumstances surrounding environmental services. The prospect of a partnership between the quantity surveyor and an environmental consultancy is seen as advantageous to both parties, combining the quantity surveyor's expertise in cost advice and the consultancy's expertise in environmental advice. Advantages have been seen to accrue to both parties when a client teams with a quantity surveyor to provide accommodation that is constructionally and environmentally sound.

CONCLUSIONS

There is no doubt at all that environmental issues will become increasingly important in the construction industry. It follows therefore that professionals with environmental knowledge and construction expertise will be in demand. This does not mean that this market belongs to the quantity surveyor by right - if the quantity surveying profession wishes to remain a proactive member of the construction team they *must* open their minds and contribute to the environmental debate.

In order to do this, however, environmental education is essential and quantity surveyors who ignore this need do so at their peril. Training is necessary on relevant legislation, management techniques and environmental issues.

With regard to the problem of demand for environmental services, this is a chicken and egg situation - which comes first, demand or supply? Shrewd marketeers may answer that if you tell people they need something, they will go out and buy it. Who thought they needed the wheel, until it was invented? No one knew what a wheel was.

Therefore, it can be concluded that, there is a niche market to be had in the cost management of environmental issues - but this market will need to be fought for.

REFERENCES

Baker, C. P. (1996). *Cost Managing the Environment, The Quantity Surveyor's Role.* Draft paper for RICS Construction & Environment Skills Panel October, 1993

Brandon, P.S. (1990). *Quantity Surveying Techniqiues - New Directions.* Blackwell Scientific, Oxford.

Bright, K. (1991). *Building a Greeener Future: environmental issues facing the construction industry.* CIOB, London.

CIRIA, (1987). *The Construction Industry and the Environment: the way forward.* Special Publication 77.

CIRIA, (1993). *Environmental Issues in Construction, Volume 2 - Technical Review.* Special Publication 94.

CIRIA, (1994). *Environmental Handbook for Building and Civil Engineering Projects, Construction phase.* Special Publication 98.

Chartered Surveyor Monthly (1994). *Survey of Chartered Surveyors in Environmental Services.* February.

Denner, J. (1994). *Contaminated Land: A framework for risk assessment.* Department of the Environment, London.

Department of the Environment (1990). *This Common Inheritance, Britain's Environmental Strategy - The first year report.* HMSO, London.

Department of Trade and Industry (1994). *An Introduction to UK Environmental Technologies.* HMSO, London.

Flanagan, R.and Norman, G.(1993). *Risk Management and Construction.* Blackwell Scientific, Oxford.

Hawkins, R. G. P. (1993). *Contaminated Land: the impact of the EPA 1990 on rural land management.* Draft report, RICS, London.

Hopkinson, L. M. (1990). *Existence Values: An economic and ethical perspective.* Imperial College, Centre for Environmental Technology, London. September.

Pasquire, C. L. & Plunkett, C. P. (1995). *An Investigation into the Application of Quantity Surveying Skills in the Cost Management of Environmental Issues.* Loughborough University of Technology, Loughborough.

Pritchard, P. (1994). *Managing Environmental Risks and Liabilities.* Business and the environment practitioner series. Technical Communications Ltd., Letchworth.

Perry, J. & Thompson, P. (1992). *Engineering Construction Risks, A guide to project risk analysis and risk management.* SERC report.

RICS (1993). *Environmental Assessments and Audits, An overview for the rural practice chartered surveyor*; November, RICS, London.

Wood, G. (1993) *Environmental statements*, Environmental Management Skills Panel, Draft paper. RICS, London.

5 Who Greens the Housing Market?

MARK BHATTI & CHRIS SARNO
Centre for Local Environmental Policies and Strategies, South Bank University, London, UK.

INTRODUCTION

Recent public opinion surveys show there is growing public concern for the environment and increasing numbers of consumers are making a conscious decision to buy environmentally friendly goods and services. If these surveys are correct, then there is much more awareness about global environmental issues than ever before. There is growing information on the environmental impact of our daily activities to help us become 'green consumers'. What is the significance of 'green consumerism' in the housing market? Is there demand for energy efficiency dwellings, and can consumers make environmentally friendly choices when buying their home? These questions raise the central issue around which this chapter is based: how can the housing market be greened? (1)

A desire to green housing, however, raises the question of precisely how market institutions and consumers are to behave in an environmentally responsible manner, and which instruments are the most appropriate to encourage them (Jacobs, 1991). These policy initiatives can range from changing the nature and direction of the market by forcing structural change in the housing system, through raising standards and stringent regulation, to economic inducements in the form of grants and cheap loans for environmental improvements. Government can also encourage environmentally responsible behaviour through advice and information campaigns in the hope that individuals and firms will take voluntary action. But will changes in behaviour by consumers be enough to green the housing market? What role can housing professionals play? Is there a stronger role for government in the form of direct intervention and tighter regulation?

More energy is used in the life time of a house than goes into its construction, but generally owners know very little about how they may choose between one dwelling and another in terms of energy efficiency, or how running costs relate to the thermal efficiency of their home. It is precisely here that the government believes market institutions such as mortgage lenders and estate agents can help. Thus in an attempt to meet targets for reduction in carbon dioxide emissions agreed at the Earth Summit in Rio de Janeiro in 1992, the UK government has

The Environmental Imapct of Land and Property Management. Edited by Yvonne Rydin
©1996, The Royal Institution of Chartered Surveyors

begun, through various policy instruments, to put some pressure on the public sector and market institutions to respond (DOE, 1994a, b, c). The precise way in which government decides to green housing clearly raises a number of issues for consumers and private institutions involved in the housing market. The main aim of this chapter is to look critically at consumer-led greening of the housing market with particular reference to energy consumption in the domestic sector (see Bhatti et al., 1994, Bhatti, 1996). The policy context for the study is the usefulness of market instruments to 'green' the housing market. The chapter is in three sections: the first sets out a typology against which the theoretical and policy discussion about why and how the housing market should be greened can be assessed. The second section reports research findings that suggest a greater role for market institutions. The final section evaluates the idea of the 'self greening' market and suggests that a stronger role for government and housing professionals is required.

GREENING THE HOUSING MARKET: THEORY AND POLICY

The UK's current environmental policy in housing has emerged out of the agreement on climate change signed at the Earth Summit in Rio (LGMB, 1992). This commits the British government to reducing carbon dioxide emissions to 1990 levels by the year 2000. There are major implications for energy use in housing as it is estimated that at least a 40% reduction in carbon dioxide is to come from cuts in energy consumption in the domestic sector (DOE, 1994c). Policies have focused action around the need to reduce the demand for energy in the home. There are basically two areas of concern: on the one hand increasing the thermal performance of the dwelling itself through raising standards, stringent regulation, and increased capital investment; on the other, changing the attitudes and living patterns of households themselves to reduce their energy use. The former is essentially about increasing energy efficiency, the latter about energy saving; policy initiatives have focused on both these areas. These policy initiatives however need to be set within a wider framework. The prevailing approach to tackling environmental problems stems from a technocentric approach: a belief in traditional institutional arrangements; new technology; a 'self greening' market; voluntary effort based on the assumption that everyone should ultimately accept responsibility for the state of the environment on the basis of the 'polluter pays principle' (see Bhatti, 1994 for a discussion). At the heart of this approach are underlying assumptions that stem from free market environmentalism.

Housing and neo-classical environmental economics

The theoretical context for the development of a 'market orientated' environmental policy is the notion of a self greening market, as proposed by neo-classical environmental economics. The central proposition is that the market place can and should take the lead in delivering sustainable development. The underpinning for this approach comes from Pearce et al. (1989) who suggest that environmental degradation results because there is no market for environmental goods and services. The market place hides the true cost of goods and services from the consumer, and the price does not reflect the full cost of the production of goods. For example, the development of a new housing estate does not include costings for a beautiful view that may be lost forever, or the increase in carbon dioxide emissions from energy use, or damage caused by the increase in car based traffic in the area.

The solution is to 'value the environment' by attaching monetary values to environmental gains and losses; what is often referred to as 'monetization'. Money can act as a measure of the degree of concern that individuals place upon a particular environmental good and can then be articulated via the market place. In this way it is possible to measure the willingness of consumers to pay for a given environmental quality and thereby work towards improving it. Pearce for example suggests,

> *'If consumers change their tastes in favour of less polluting products and against more polluting forces/ market forces will lead to a change in the "pollution content" of the final product and services' (p.155).*

The 'green consumer' has often been seen by some environmentalists as the driving force behind saving the planet (Elkington and Hailes, 1988). Individual preferences of the purchaser are seen to be the key in forcing producers to 'clean up their act', to promote environmentally friendly products, and for everyone to accept responsibility for the environment. If 'green demand' can be identified producers can actively target the green consumer as a market niche to be cultivated. The outcome of the neo-classical approach is that the market acts as the central mechanism for tackling major environmental problems, with the green consumer leading the way.

But some environmentalists in the neo-classical camp do recognise that left to itself the market does not work perfectly, and can produce waste and misuse resources. Pearce, for example, argues that there have to be market based incentives in the form of pollution charges to raise the price of products that have a negative environmental impact, or tax reductions for those products that have a minimal impact. The charge or reduction in price allows the consumer to make their demand for environmentally friendly goods effective. Producers can decide how best to respond, either by investing in pollution abatement technology, or switching to products that cause less pollution. There is thus a role for government in creating market incentives via taxation measures, as well as offering information and advice to consumers so they become environmentally aware citizens. The relevance of Pearce is that his argument for market intervention is clearly acceptable to those who believe in a 'free market'. Thus over the last 15 years the government has shifted the emphasis towards market orientated solutions to social, economic and environmental problems. In housing this involves giving the consumer greater freedom and choice to buy energy saving products, and providing advice and information on energy conservation.

Policy implications

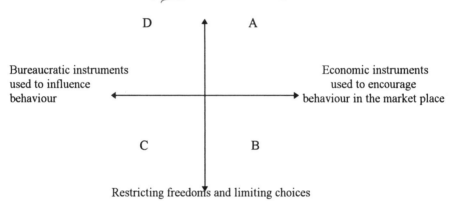

Enhancing freedom of choice and responsibilities

D   A

Bureaucratic instruments
used to influence
behaviour

Economic instruments
used to encourage
behaviour in the market place

C   B

Restricting freedoms and limiting choices

Figure 1 Analysing the impact of different instruments on indiviuals and firms
Source: Adapted from Young, 1994

Young's typology in figure 1 outlines the use and impact of different bureaucratic or market instruments which could be used to green the housing market. Along the horizontal axis we have a market orientated approach towards the right hand side, suggesting that consumers and producers will only respond to supply and demand based indicators with minimal state intervention. On the left hand side of figure 1, a more interventionist approach calls for regulation and bureaucratic control of the housing system. This can be linked to the vertical axis to see whether these two instruments limit choices for consumers and firms, or whether they enhance freedom of choice.

Young's typology helps us to examine the range of measures that could be used by government to help reduce energy use and therefore carbon dioxide emissions in the housing sector.

In boxes B and C the idea is to limit choices and attempt to control the worst behaviour. For example, VAT on fuel falls into box B as it has been presented as a 'green tax' whereby everybody has to pay, and it is argued that individuals will use less fuel. In box C we are likely to see regulation in the form of minimum standards for new products or rules to allow individuals to buy only environmentally friendly goods and services. Thermal efficiency has been a part of Building Regulations since the 1970s and standards have progressively been raised. Under the new Building Regulations (applicable from 1 July 1995) energy efficiency standards have been raised further. For the very first time builders will have to provide an energy label for each new dwelling. The overall energy rating will be assessed under the Standard Assessment Procedure (SAP). SAP is the government's recognised energy labelling system, and gives a rating from 1 (low) to 100 (high). The SAP, as an energy label, is designed to inform householders of the energy efficiency of the house and the running cost of the dwelling (BRECSU, 1994). It takes into account the thermal insulation of the building fabric, effectiveness of the heating system, solar gain, ventilation, and the price of fuel. For new dwellings and conversions where new units are created, the SAP target will be set at a minimum of 60, increasing with the floor area of the dwelling. A typical 3 bed house for example will have an SAP target of 80. In many respects, the incorporation of an SAP target into the Building Regulations represents a major step forward. However there are criticisms of SAP which seem to suggest that it may not be as effective as government hopes (Bhatti, 1996, Shove, 1995).

The prevailing approach to tackling environmental problems is highlighted in boxes A and D, so that government action enables consumers and firms to respond voluntarily, basing decisions on their individual circumstances. Economic instruments in box A can take the form of financial incentives. For example, tax breaks such as VAT exemption on new homes that are more energy efficient; this encourages responsible behaviour. In box D, advice and information on greener practices and products could be provided. Vital information such as energy rating of new dwellings can help consumers choose between different homes. Building firms can be made aware of the consequences of their production methods so that they are better to change practices.

The government's approach towards environmental issues over the last decade falls into boxes A and D because the majority of initiatives to date have suggested that it does not believe in direct intervention.

The main policy initiatives so far are based on the assumption that the market place is to take the lead in promoting energy efficiency measures along with voluntary effort by developers and housing consumers. 'This Common Inheritance' (DOE, 1990) announced this approach by arguing that initiatives on energy conservation were designed to inform consumers and 'encourage them to take action'. John Wakeham, then Energy Secretary, made it clear in a speech to the House of Commons,

> 'The Government's view is that the market remains incomparably the best and most powerful mechanism available for inducing improvement in energy efficiency, which remain dependant on the decisions taken by millions of individuals' (5 September 1990).

Thus no energy efficiency targets for the housing stock have been set as it is believed this would interfere with the choices and decisions of individual householders. The government's commitment to a voluntary approach involves increasing the amount of information flowing through to consumers in the hope that they will make environmentally-informed decisions about buying energy efficient homes, investing in energy conservation and reducing their fuel use. Hence a market led approach with minimal regulation underpins much of recent environmental policy in housing in the UK.

A number of issues for the housing market follow on from this approach to sustainable development that are central to this study. If green consumerism is a key force for sustainable development, then what is the nature of this 'green demand', and what are the implications for the future of the private housing market? For sustainable development to be achieved solely through the market, then green consumers are key players; we need to understand how they behave in the housing market. The second section reports from housing users themselves about the significance of energy saving features when purchasing new homes.

Thus new regulations in the form of energy rating of new homes (SAP), and demands for eco-labelling from the EU, may begin to have an effect on demand for energy efficient homes. Thus the government hopes that increasing the information base from which consumers can make informed decisions on house purchases, could also make a significant difference in the market. In a preliminary way the research examines the impact of various policy instruments on attempts to green the market from the perspective of housing consumers and assesses their response to these. Various institutions in the market also need to consider precisely how 'green demand' can be managed. The study reported in this chapter looks at the impact of these developments on house purchase decisions. What is the place of energy efficiency of the dwelling in relation to other locational factors? Are consumers willing to pay a premium for energy efficient housing? What kinds of information would help buyers to make informed decisions? Should housing professionals target the 'green' housing consumer as a 'niche' market to be cultivated? For the purposes of this chapter the definition of 'green demand' is equated with demand for energy saving even though there may (or may not) be demand for other services and products in the home (e.g. CFC free fridges). The study takes this narrow definition of green demand and through a survey methodology examines attitudes and decisions of house purchasers in a specific sub-market, that is, households moving into new homes (2).

DEMAND FOR NEW HOMES AND GREENING THE HOUSING MARKET

It is now widely accepted that improving energy efficiency of the existing stock must be the main priority as new stock consists of only 1-2% of total dwellings in any one year (Gibson, 1994). So why look at new homes? The new homes sub-market is where we expect to find signs of 'green demand' as defined above.

Standards of thermal performance have been improving; since 1990 there have been significant changes to building regulations specifically designed to reduce energy consumption in the home (3). In recent years if households demand energy efficient homes then they are more likely to find the product they want in the new homes market. So one question we focus on is the extent to which energy saving features are a factor in the decision to buy a new home.

Whilst it is true that, in the short term at least, significant reductions in carbon dioxide in the domestic sector can only be achieved by concentrating on the existing inefficient homes, in the long term, improving thermal performance of new dwellings can have a major impact. This is because firstly, new dwellings can set a precedent that could be transferred to existing homes. Energy rating is an example of this and we discuss it below. Secondly, there is a complex relationship between new and second hand markets, but innovation and new techniques can filter through from the former to the latter.

Moreover the experiences of households in new homes can permeate the second hand market in relation to comfort and design more generally. Thirdly, how new homes are marketed is an important factor in the growth of green demand in housing. Thus some developers and estate agents are highlighting the benefits of energy efficiency of their product, largely to differentiate it from the second hand market. Even so, it would be wrong to overstate the extent to which this is happening; it may be a feature of a stagnant housing market and quickly disappear when house prices pick up. Looking only at the new homes sub-market we cannot generalise from our survey, but it does allow us to study the nature and impact, if any, of green demand in a controlled environment.

Households and new homes

Households moving into new homes have particular characteristics that are important to highlight for the purposes of our study. Government research into new homes (DOE, 1993) shows that purchasers are likely to be younger professional/managerial small families or without children, living in detached houses (DOE, 1993 p.34). This is also reflected in our survey in that 60% of respondents were either single or two adult households, with 31% having children under the age of 18. The age profile is also very similar in that nearly three quarters of the respondents were aged between 26 and 50. The dominant dwelling type is the detached (51%) with 21% living in semi-detached houses. Even though we did not enquire into the socio-economic profile of households there is no reason to suggest that our respondents differ greatly from that found in other surveys (DOE, 1993 p.36-37). There is high turnover in this sub-market, and 50% of households move less than five miles to buy a new home.

Previous research on attitudes to energy saving (DOE, 1991, PiP, 1994) has suggested that though owner occupiers are concerned about the environmental problems, there are significant barriers to overcome when it comes to action by individual households. Much of this research has focused on higher income households, often described as the 'fuel rich'. The fuel rich are those households who could invest in energy saving features (or purchase new energy efficient homes) but choose to waste energy instead. It is likely that our sample could be described as the fuel rich, as the new homes market is dominated by richer households. However it is also this group who could demand and pay for greener products and thereby stimulate the market place.

BUYING A HOME

It is generally accepted amongst estate agents that the first law of house buying is location of the dwelling in relation to work and other facilities. However in the 1990s things may be changing; we asked respondents to place their top three priorities when searching for their home.

Our survey found that only a third (31%) of households put 'near work' as their number 1 priority; 22% of respondents put 'cleaner neighbourhood' and 12% 'more greenery' as their first priority. If we look at second order priorities then 18% of households wanted 'more greenery'; 'away from traffic' and 'cleaner neighbourhood' came equal second with 17%.

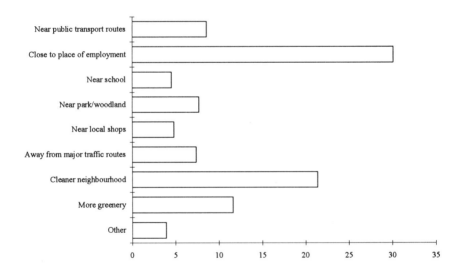

Figure 1 Decision to locate in an area

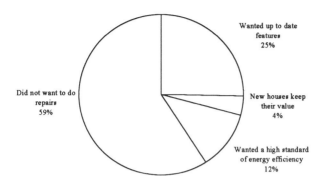

Figure 2 Why did you want to buy a new house?

This suggests that local environmental quality is beginning to play a significant role for more and more households. In fact this is backed up by the 1995 TSB House Hunters Shopping List survey which suggests that buyers aim to pay less for location, and more for houses in a good condition with gardens and in 'cleaner' neighbourhoods.

NEW HOMES AND THE 'GREEN' CONSUMER

If demand for energy efficient homes is becoming a feature of the housing market then the most likely place for it to be present is in the new homes market. This is mainly because new homes have much higher levels of energy efficiency and offer increased comfort at lower running costs.

Even so, our survey found that the major reason for buying new homes is that respondents did not want to do repairs or maintenance (59%); a quarter wanted up to date features; and only 12% put 'high standard of energy efficiency' as their top priority. It does not appear from our survey that the decision to buy a new home is based on significant demand for energy efficiency. However 37% of households did consider energy efficiency as a second order priority, which seems to suggest that there is some demand.

Green ideas may not be a motivating factor behind choice of new homes; however the effect of choosing a new home as opposed to an older one (other things being equal) is the reduction in energy use and therefore carbon dioxide emissions of the key issues in this area is that the extent to which 'green' demand is 'out there' searching for green products or the extent to which it has to be cultivated. The survey reveals that nearly half of our respondents (48%) enquired about the energy efficiency of the house they were going to purchase. However the majority (53%) also said the estate agent did not highlight any energy saving features of the new home, and only 36% received any energy information, despite the fact new homes built in 1994 have much higher standards of energy efficiency.

With the media coverage of VAT on fuel many households wish to reduce their fuel bills. The overwhelming majority (85%) of our respondents said they were concerned about conserving fuel; of these, 77% turned down heating, and 40% turned lights off when not needed but only a few (9%) heated only one room. Most respondents (58%) said that their present house could be made more energy efficient, which seems to suggest that consumers believe that more could be done. Thus 61 mentioned triple glazing, and 55 mentioned improved draught proofing. Many of the comments related to poor building design, use of inferior materials, and poor workmanship; some owners complained that their new homes suffered from poorly fitting doors and windows. A few felt stairs should be sealed off so that heat does not escape from living rooms.

These findings demonstrate that the demand for energy saving advice and information is rising, and yet estate agents are unable, or do not want, to respond. It is probably the case that estate agents will respond under two circumstances: if consumers are willing to pay more for energy efficient homes; or they have to because of government regulations. We examine these below.

We wanted to know how much more new owners would be willing to pay for an energy efficient house if they could get their money back over a specified time period. The results are mixed. 36% of households were unwilling to pay more, but exactly the same percentage were willing to pay 5% more if they could get their money back within two years. In general respondents were willing to pay more only if they recoup the money quickly. However, 27% opted to pay 7.5%-20% more and with pay back periods between 3-10 years.

This cluster appear to be the 'greenest' in that proportionately more of this group enquired about energy saving (58% as against the survey average of 48%); 43% were willing to raise their offer price for a property with high energy efficiency standards compared with the survey average of 28%. The survey does suggest therefore that a specific group are interested in energy efficiency and are willing to pay more for it. The results however are inconclusive and more qualitative research focusing on this type of group needs to be done.

ATTITUDES TO ENERGY RATING

When households go through the process of buying they receive a range of information which may influence the choice of their future home. As mentioned above estate agents did not generally respond to demands for information on energy saving measures. We asked specifically about demand for energy labelling information.

The most significant finding from our survey is that when buying a house most respondents (87%) wanted to know the energy rating of their new home (called

SAP; see note 3). Thus there appears to be a strong desire for energy labelling information.

We next asked who should provide that information. 42% of the sample believed that estate agents should provide them with the energy rating of their house. The 'surveyor' and the 'local council' came out equal second at 21%. Indeed 59% of respondents wanted the SAP rating at the same time as they get details of a house from the estate agent.

The implications of these findings are that SAP ratings need to be carried out at the same time as details of the house are being compiled by the estate agent (this happens in Denmark for example), thus incorporating the SAP rating into the surveyor's/valuer's report may not be appropriate.

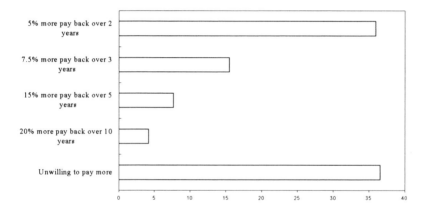

Figure 3 How much more would you be willing to pay for a highly insulated house?

There is a role here for surveyors to check that the SAP rating is correct, and to adjust if necessary. In effect the surveyor would be acting as a second check.

From energy rating information some households may begin to consider energy efficiency as an important factor when considering which property to buy, and how much to pay. For example, 83% said they would reduce the offer price if the house they wished to buy had a low energy rating. However, far fewer (28%) said they would be willing to pay more for a higher rating. The majority (84%) wanted energy rating certificates to be made compulsory for new homes, and 77% wanted them for all homes. If SAP is made compulsory for all dwellings, it will have a major impact on the housing market; the marketing and exchange of dwellings will also have to adapt.

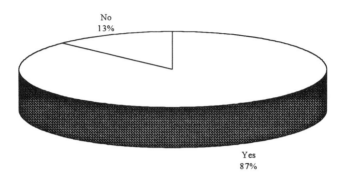

Figure 4 Would you want to know the SAP rating of your house?

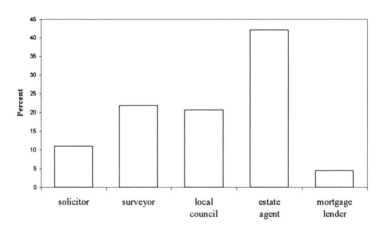

Figure 5 Who would you like to provide you with this information?

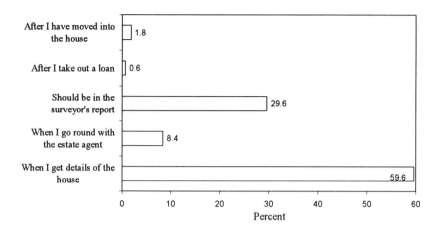

Figure 6 When would you prefer this information to be provided?

So how should market institutions in general, and estate agents in particular respond? Home buyers rely on estate agents for information and, according to our survey, this will include energy labelling. With over a million transactions every year there is clearly an opportunity for agencies to incorporate an energy dimension into the exchange process.

CONCLUSIONS

This chapter has critically assessed initiatives in the private housing market, aimed at improving the energy efficiency of the stock within the context of wider environmental policy. Drawing on Young's typology, the research suggests that there are limits to the market-orientated approach. One conclusion in that market institutions need to do much more in the way of encouraging the demand for energy efficient homes. Government can help by enabling bureaucratic instruments to engage positively with market based initiatives.

The policy document 'This Common Inheritance' effectively placed home energy efficiency within the context of the new environmental agenda and should have helped to focus government initiatives in the market place. However there is now real concern that the target of a 40% cut in carbon dioxide emissions from the housing sector will not be achieved. This is mainly because the government has sought to pursue policy through a voluntary and market led approach, which our research suggests is fairly limited. Even though there are some encouraging signs, in assessing the impact of environmental policy on the private housing market there is little evidence of sustained 'greening'.

The survey does not reveal 'green' demand in the housing market to any great extent. The energy efficiency of the dwelling does not as yet figure prominently in the decision to opt for a new home. However, there is a large group who, in the process of looking for new homes, demand energy advice and information, and a significant minority who opt for new homes because of energy saving features. This group has to be cultivated. Because of growing public concern for the environment, and increased fuel costs, it may only be a matter of time before energy use becomes a key feature of the housing market. In many ways this provides new opportunities for surveyors and valuers and a chance to gain new skills.

The current lack of demand could be due to inadequate responses from market institutions themselves.

Firstly, very few estate agents appear to be marketing new homes in such a way to allow potential buyers to consider energy efficiency as one important element in their choice. Thus, over half of the respondents said that energy saving features were not highlighted; one of the major advantages of a new home, the increased standard of energy efficiency, is not generally being marketed.

Secondly, estate agents and building societies need to consider the extent to which consumers will take low SAP ratings into account in their search for properties. According to our research low ratings could lead to a downward pressure on house prices. However, this could also act as an incentive for vendors to invest in energy efficiency measures in order to improve SAP ratings. Building societies need to develop new products and/or strategies, for example soft loans for energy efficiency measures; or retaining a small part of the loan until energy related works are completed. Again estate agents have a role to play in alerting buyers to new loan products that may be available.

Finally, previous research funded by the RICS shows that awareness of environmental issues is quite low amongst building societies and estate agents. Indeed our research suggests that estate agents continue to act as a brake; they need to develop 'green' marketing strategies for the products they are trying to sell. The inclusion of SAP ratings will have an impact on the housing market, and this requires further research. Our study shows that there is significant demand for energy labelling; now it is up to estate agents, surveyors, and building societies to respond imaginatively.

ACKNOWLEDGEMENTS

We would like to acknowledge funding from the RICS for this research. Thanks also to the NHBC for providing a sample from which the survey was carried out.

NOTES

1. While green in lower case is used to describe policies and practices displaying a level of concern about the environment, Green with a capital G refers to a specific position put forward by the Green Party or Greenpeace.
2. The survey was conducted over the summer of 1995, with a 2% random sample drawn from NHBC registrations over 1994 (92,000). 1339 questionnaires were sent out and 336 received for analysis achieving a response rate of 25%.
3. Our sample was taken for dwellings that received a NHBC certificate during 1994 so houses built in 1993/94 will come under the 1990 Building Regulations (revised in 1992). The regulations of this period are stringent in relation to energy efficiency in that targets have been set that must be met for every new dwelling. The sample contains the most up to date standards of energy efficiency prior to 1 July 1995.

REFERENCES

Bhatti, M. (1994). Environmental futures and the housing question. In Bhatti, M., Brook, and Gibson, J. (eds) *Housing and Environment: A New Agenda*. Chartered Institute of Housing, Coventry.

Bhatti, M., Brook, J. and Gibson, M. (eds) (1994*). Housing and Environment: A New Agenda*. Chartered Institute of Housing, Coventry.

Bhatti, M. (1996). Housing and environmental policy in the UK. *Policy and Politics* April.

BRECSU (1994). *The Government's Standard Assessment Procedure for Energy Rating of Dwellings*. BRE, Watford.

DOE (1990). *This Common Inheritance*. HMSO, London.

DOE (1991). *Attitudes to Energy Conservation in the Home*. HMSO, London.

DOE (1992). *The UK Environment*. HMSO, London.

DOE (1993). *New Homes for Homeowners*. HMSO, London.

DOE (1994a). *The Strategy for Sustainable Development*. London: HMSO

DOE (1994b). *Energy Efficiency in Council Housing: Guidance for Local Authorities*.DoE, London.

DOE (1994c). *Climate Change: The UK Programme*. HMSO, London.

Elkington, J. and Hailes, J. (1998). *The Green Consumer Guide*. Gollanz, London.

Gibson, M. (1994).The greening of housing policy. In Bhatti, M., Brooke, J. and Gibson, M. (eds.) *Housing the Environment: A New Agenda*. Chartered Institute of Housing, Coventry

Jacobs, M. (1992). *The Green Economy*. Pluto Press, London.

LGMB (1992). *The Earth Summit Rio 92, Agenda 21 - A Guide for Local Authorities in the UK.* LGMB, Luton.

Pearce, D., Markandya, A. and Barbier E.B. (1989). *Blueprint for a Green Economy.* Earthscan, London.

Projects in Partnership (PiP) (1994). *Breaking Through the Barriers to Energy Saving in the Home.* PiP, London.

Rydin, Y. (1994). The greening of the housing market. In Bhatti, M., Brook, J. and Gibson, M. (eds) *Housing and Environment: A New Agenda.* Chartered Institute of Housing, Coventry.

Shove, E. (1995). Constructing regulations and regulating construction: the practicalities of environmental policy. In Gray, T. (ed) *Environmental Policy in the 1990's.* MacMillan Press, Basingstoke.

Young, S. (1994). An Agenda 21 Strategy for the UK. *Environmental Politics* **3**(3): 325-334.

6 The Effect of the New Energy Conservation Regulations on the Design, Cost and Energy Consumption of Refurbished Buildings

SANDRA JANE DIXON
University of Northumbria at Newcastle, Newcastle-upon-Tyne, UK

SUMMARY

Building Regulations governing thermal insulation standards were first introduced in 1979 and were revised and extended in 1985, 1991 and most recently in 1995. By the year 2000 some 70% of buildings at that date will have been built before any significant fuel conservation regulations. The potential for energy savings in the sector of existing buildings is therefore substantial.

Parts of the Building Regulations (Amendment) Regulations 1994 (SI 1850) came into force on 1 July 1995. For the first time, the Regulations allow for the requirements relating to the conservation of fuel and power to be applied to certain building work on *existing* buildings. This paper aims to examine the effects these changes will have on the design, cost and energy consumption of refurbished buildings falling within the scope of controlled work. The scope of the new Regulations is examined with regard to the type of building uses which are controlled and the extent to which the new requirements should be applied.

A case study is used to illustrate the work which is required in upgrading existing buildings to meet the requirements of the new Approved Document L; the additional costs of improving the energy efficiency of altered buildings; and the amount of energy saved by the improvements.

The paper discusses how these changes to Building Regulations fit in with the overall scheme implemented by government to reduce emissions of carbon dioxide, 'The Climate Change Programme' (CCP), and the challenge which they pose to the UK construction industry.

BACKGROUND

Energy efficiency in buildings has assumed greater environmental significance for

The Environmental Impact of Land and Property Management, Edited by Yvonne Rydin
© 1996, The Royal Institution of Chartered Surveyors

the following two reasons.

Firstly, it is one of the instruments which government is using to reduce our emissions of greenhouse gases and so meet our commitment to the United Nations Framework Convention on Climate Change which was signed at the 'Earth Summit' in Rio de Janeiro, June 1992.

Secondly, buildings use large amounts of energy. In 1991 43% of final energy demand in the UK was accountable to domestic, public and commercial buildings which resulted in half of all national carbon dioxide emissions (the greenhouse gas which is increasing in proportion faster than any other) (Department of Trade & Industry, 1994).

In October 1992, the government projected that we need to reduce emissions by some 10 million tonnes of carbon (MtC) to enable us to meet our Rio commitment and return to 1990 levels of carbon dioxide and other greenhouse gases by 2000 (Department of the Environment, 1994a). The government's 'Climate Change Programme' was announced in January 1994, it aimed to reduce the amount of fossil fuels used in the UK through efficient use of energy and by changing the *types* of fuel used in the production of electricity, and so decrease the emissions of carbon dioxide.

Specifically, the programme included:

- Strengthening the Energy Efficiency Office's (EEO) programme of advice and information aimed at business;
- The establishment of the Energy Saving Trust (EST) to provide financial incentives to energy efficiency in the domestic and small business sectors;
- The introduction of VAT on fuel as a fiscal incentive;
- An increase in the objectives for capacity in place by 2000 for both renewable energy and combined cycle gas turbines (CCGT) for electricity generation; and
- Improving the Building Regulations for the conservation of fuel and power (Department of the Environment 1994a).

The target figures were revised in March 1995 and it is now expected that emissions reductions of 17-25 MtC will be achieved by 2000. This revision has been necessary to take into account the changes in the supply-side of electricity generation where the take-up of combined cycle gas turbines (CCGT) has been much greater than expected (Department of Trade & Industry, 1995).

This is in contrast to the disappointing results of the other measures, which mainly target energy efficiency in *buildings*, where,

- severe problems have arisen in the funding arrangements, especially for the Energy Saving Trust,
- the take up of EEO initiatives by building owners, managers and developers and has been slow, and,
- political problems have been experienced, particularly in the introduction of VAT on fuel prices.

Table 1 Government targets for reductions in CO_2 emissions (January 1994)

Sector	Expected reduction in emissions by 2000, MtC
Energy Consumption at Home	4
introduction of VAT on domestic fuel use	
new Energy Saving Trust	
energy efficiency advice	
'Helping the earth begins at home'	
eco-labelling	
EC SAVE programme (standards for household appliances)	
revision of Building Regulations to strengthen energy efficiency requirements	
Energy Consumption by Business	2.5
energy efficiency advice	
Making a Corporate Commitment	
Best Practice Programme	
Regional Energy Efficiency Offices	
Energy Management Assistance Scheme	
Energy Saving Trust schemes for small businesses	
Energy Design Advice Scheme	
possible EC SAVE scheme (standards for office machinery)	
Revision of Building Regulations to strengthen energy efficiency requirements	
Energy Consumption in the Public Sector	1
targets for central and local government and public sector bodies	
Transport	2.5
increases in road fuel duties	
TOTAL	10

Note: savings in the electricity generating sector have been allocated to final users (including encouragement of renewable energies and Combined Heat and Power plants). This table is intended as a summary of the key measures. Allowance has been made for overlap between some of the individual programmes.
Source: DoE, 1994a

Of the total amount of CO_2 to be saved by the initiatives outlined in Table 1, the improvements to Building Regulations were expected to save only 0.25 MtC (Department of the Environment, 1994a). The 1995 revisions to Approved Document L have provided a substantial improvement compared to the 1992 edition of the Document. However, they were introduced one year later than anticipated and did not include all of the intended requirements.

Improving the energy efficiency of *existing* buildings is one of the main vehicles for the construction industry to contribute to the Climate Change Programme. The success of the Energy Efficiency Office schemes relies on the commitment of the funding bodies and the willingness of property developers to implement energy efficient technology into building projects; so far there is little evidence of this happening on a voluntary basis. The success of the new Building Regulations relating to energy conservation relies on the level of environmental education within the construction industry, and, the workmanship and quality control on site.

The changes to the Building Regulations

The need to reduce the amount of energy consumed in existing buildings was recognised by the House of Commons Select Committee on the Environment in their fourth report which was published in 1993; it made 63 recommendations to the government. They stated:

> *'We strongly recommend that Building Regulations be revised in order to raise standards of energy efficiency for new build properties. We also recommend that they be extended to major refurbishment and repair of existing properties, and that such material alterations be clearly defined.'*

The Royal Institution of Chartered Surveyors agreed with this recommendation, stating in a November 1993 press release;

> *'If Britain is to become an energy efficient society it is essential that the Government finds the political will to finance the latest recommendations from the House of Commons Environment Committee'.*

The Committee's recommendations were
- to promote and standardise energy rating of dwellings,
- allow designers flexibility whilst discouraging 'trade-offs' between 'U' values,
- encourage improvements in workmanship to avoid technical risks,
- raise energy efficiency standards in the Regulations,
- extend the scope of Approved Document L to cover material alterations and changes of use,
- encourage the use of low energy lighting, and
- for the government to work with Professional Institutions to incorporate energy efficiency into educational and training courses for building design and management skills to discourage unnecessary air conditioning and mechanical ventilation in buildings.

The 1995 edition of Approved Document L and its supporting document 'Thermal Insulation Avoiding Risks' cover all but the latter of these recommendations.

The Department of the Environment has chosen to introduce measures in the 1995 edition of Approved Document L which it considers to be cost-effective, but which do not introduce unnecessary technical risk. It estimated that construction costs for *new* buildings will increase by 1-2% and pay for themselves in 8-15 years and that energy performance will be improved by 25-35% compared with the 1990 standards (National Audit Office, 1994). However, similar claims were made when the 1990 standards were brought into effect. Research at John Moores University, Liverpool found that houses built since the 1990 amendment are saving an average of 6.3% of energy consumption, not the 20% reduction anticipated by the government (House of Commons Environment Committee, 1993).

Changes to the requirements which affect existing buildings

This paper will only examine the sections of the new Regulations which affect alterations and refurbishment of property.

Regulation 6.1(a) 'Requirements relating to material change of use' has been revised to include L1 'Conservation of Fuel and Power' in the list of requirements. The Approved Document is quite specific on the work which is required to show compliance when carrying out material alterations or changes of use. Regulations 3 and 5 define what type of work falls into these categories.

Regulation 14A has been added. It requires all new dwellings, and those created through change of use of a property, to have an energy rating carried out. This rating should be carried out using the Standard Assessment Procedure (SAP).

The requirements relating to refurbishment of properties are more specific in Approved Document L than any other. The upgrading of the insulation levels when carrying out material alterations and change of use work *'will depend upon the circumstances in each case'*. Where it is considered that the circumstances are appropriate, the document gives specific guidance.

The changes to the Regulations are significant to building surveying organisations as 39% of their work involves refurbishment, repair and maintenance of properties (RICS, 1990). Work to existing buildings is increasing its importance in the industry, in the year ending September 1994 the total value of repair and maintenance work in the UK was 5% higher, at about £24 billion, than the previous twelve months (Department of the Environment, 1995). Research was carried out during 1994 by University College London into what factors affect the decision to refurbish buildings for change of use. They surveyed groups who are concerned with the strategic, economic and technical viability options of change of use developments. The groups of decision makers were investors, producers, marketing, developers, regulators and corporate users. When asked which physical characteristics of a building are positive attributes for changing its use, building character and period features had high levels of response. It is therefore fair to assume that the types of buildings which are most likely to be affected by the changes to the Building Regulations are old, traditionally constructed premises with some features considered worthy of preservation.

AMOUNT OF WORK AFFECTED BY THE CHANGES

A survey has been conducted of Local Authority Building Control (LABC) Departments in England and Wales. Eighty three Authorities were questioned regarding the percentage of Building Regulations applications which involve materially changing the use of buildings and 33 useful responses were received. The survey found that 4% of all Building Control applications fall into this category, of which three-quarters are to change the use of the building to a dwelling or flat (hence needing a SAP survey to be carried out).

It is virtually impossible to gain a national picture of the actual number of dwelling units which are created through change of use applications as a single application can cover several properties. The DoE's 'Housing and Construction Statistics' only report the total number completed per year and they do not hold records which distinguish between new units and those created through change of use. Therefore, the amount of energy saved nationally by these changes is difficult to quantify.

ALTERATION WORK WHICH IS NOT CONTROLLED BY THE REGULATIONS

It is worth noting at this point the type of work which is *not* controlled by the Regulations to suggest where they can be tightened up in the future.

A refurbishment of an existing office building may constitute a material alteration to the property, but because artificial lighting is not a 'controlled service' as defined by Regulation 2, then there will be no requirement to improve the energy efficiency of the lighting installation should it be being replaced.

Similarly, a building which is being changed to an office use is not controlled under Regulation 5 as the only non-domestic uses covered are hotels, boarding houses, institutional and public buildings. It is reported that 4% of existing office floor space in England and Wales has been created through conversion from another use (Centre for Configural Studies, 1994) and office refurbishment is increasing its share of the market. In 1993 the RIBA reported that between the final quarters of 1992 and 1993, refurbishment of offices increased to 39% from 27%. This indicates that a significant area of building work is not being covered by the changes to the Building Regulations as offices have been identified as one of the principal energy users where there is considerable scope for reducing energy consumption (Department of the Environment, 1992).

Other material alterations to buildings which are not covered by the recent changes are when windows are being replaced. This work will only be covered by the Regulations if it involves a structural alteration to the fabric, i.e. when the opening is being widened. If the replacement windows are of the same width as

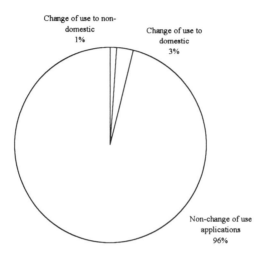

Figure 1 Percentage of Building Regulations applications which are effected by the changes to the Regulations

the existing windows then this work is not controlled and it would be possible to install glazing which does not meet the minimum standard for new work, 'U' value of 3.3 W/m²K.

ENERGY CONSCIOUS REFURBISHMENT

In most instances, designers will carry out the minimum amount of work required to comply with Approved Document L unless their client has specifically requested an energy efficient design. It can be argued that designing to minimum standards using the 'elemental approach', and following the guidance on permitted areas of glazing, is not the most effective method of providing an energy efficient design, especially when the efficiency of the building services and the use of passive solar energy are ignored

A successful refurbishment scheme in terms of energy conservation can be achieved by taking a *holistic* approach to the problem. In the preliminary stages of the project, the building(s) must be carefully inspected to determine the many factors which will work in favour of, or against, an energy efficient design. The 'U' values of the external elements are the most obvious, and perhaps the easiest type of data to be collected. The surveyor should also record several other factors to complete the 'audit' of the existing buildings performance. The guidance contained in the RICS publication - 'Energy Appraisal of existing Buildings - A Handbook for Surveyors' - is invaluable in this respect. Once the existing building's energy performance is quantified, the specification for the alteration work should seek to capitalise on the positive aspects and upgrade areas of the building and its services which are less efficient. In many cases this should lead to making maximum use of the *volume* of the building to avoid excessive unheated spaces which should have enclosing walls and floors with 'U' values of 0.6 W/m^2K.

The requirements of Approved Document L are quite basic, and, at the present time, deal only with the conservation of energy for *heating* buildings. Upgrading fabric insulation levels, reducing air infiltration and improving the controls over heating services are energy conservation techniques which are well documented by the BRE and BRECSU in Information Papers, Digests, Reports, Good Practice Case Studies, and the Best Practice Programme. It is therefore expected that they should be understood by the industry and implemented on site without too many technical difficulties. This remains to be seen in the coming months as the industry tackles their implementation.

Fabric insulation is the most widely understood element of energy conservation in the UK, however, it is only one of a series of measures which can be employed to improve the energy performance of property. This is demonstrated in the case study below. This example is intended to indicate how energy efficient designs can be produced by taking into account the holistic approach outlined above.

CASE STUDY - CHANGE OF USE FROM INSTITUTIONAL BUILDING TO DWELLING

The change of use took place in 1993 from a Health Authority building to three dwellings. Originally built in 1873, the detached building formed part of a hospital complex. It is constructed of cavity walls (225mm outer leaf, 50mm cavity and 110mm inner leaf) with suspended timber floors and dual pitched, king post roof construction. The windows are timber-framed, sliding sash type with stone cills, lintels and jambs. None of the external elements was insulated.

One of the dwelling units created by the change of use will be examined here. The floor area of the dwelling is $51m^2$, it is single storey with extensive single glazed areas facing south east and south west. During the conversion the following alteration work was undertaken:

* the ground floor covering was substantially replaced with new boarding,
* the existing slate roof covering was replaced with concrete tiles,
* new single-glazed window and door openings were formed (the original window openings were not altered due to planning restrictions),

- each dwelling was supplied with a new space and water heating system,
- the plaster was renewed on the internal face of the external walls, and
- a suspended ceiling was installed in the living areas to reduce the floor to ceiling height.

Under the Regulations in force at the time of the conversion, there were no requirements to improve the energy efficiency of the building although the developer did insulate pipework and laid 150mm of insulation over the loft floor and provided roof ventilation.

In this project, the fabric of the building would be relatively easy to upgrade. The roof and timber floor were accessible during refurbishment to enable the insertion of insulation, however, the roof insulation which was provided was laid at the loft floor level rather than above the suspended ceiling which led to a larger volume than necessary being heated. The roof had overhanging eaves to provide adequate ventilation openings. The building had some new windows and doors inserted and this would enable draught-stripping and double-glazing to be specified. The walls were being internally renovated and this allows for insulated dry-lining to be included.

The upgrading of the insulation levels in this building would not induce any technical risks to the design and most of the work uses tried and tested techniques. Condensation can be avoided by adequate roof ventilation and the indoor air quality is kept at an acceptable level by the provision of mechanical ventilation to the kitchen and bathroom. The table overleaf details the additional work which would be required if the property had been converted under the control of the 1995 edition of Approved Document L.

All costs have been calculated excluding professional fees and VAT but including overheads and profit. The conversion costs in 1st quarter 1993 were £401.78/m^2 floor area. At today prices (3rd quarter 1995) these would be £409.13/m^2. The cost of the work if carried out today, following the introduction of the new Regulations, would be £ 438.84/m^2. This represents an increased capital cost of 7.2% (This would have been considerably lower if cavity wall insulation was specified rather than dry-lining. The Approved Document does not suggest this option although it is unlikely that any Building Control Authority would refuse such a change as it complies with the 'spirit' of the Regulations.) Figure 2 indicates annual fuel bill savings of £176.62 (these were calculated using the SAP software with natural gas being the main heating fuel for both calculations), the simple payback period on the additional work to comply with the 1994 Regulations would therefore be 8.6 years.

Table 2 Additional work required by the 1994 Building Regulations for change of use work

Specification in 1993 (under the control of the 1991 Building Regulations)	Specification to comply with 1994 edition of Building Regulations	Additional cost (£) /m^2 floor area
Walls - 'U' value unchanged at 1.2W/m^{2}K	Walls - 'U' value changed to 0.43W/m^{2}K by insulated drylining	14.95
Floor - 'U' value unchanged at 0.76W/m^{2}K	Floor - 'U' value = 0.35 W/m^{2}K 60mm of mineral wool quilt (Table A10)	4.75
Roof - 'U' value changed to 0.3 W/m^{2}K	Roof - 'U' value = 0.25W/m^{2}K 200mm of quilt insulation (Table A2) and therefore requires eaves ventilation	3.84
Windows and door - 'U' value to remain at 4.7 W/m^{2}K	Windows and door - 'U' value changed to 3.0W/m^{2}K (Table 2) for new openings and existing windows remain at 4.7W/m^2/K	1.76
Draughtstripping - none	Draughtstripping - all new windows, doors to be draughtstripped	included in figure above
Space and hot water heating not controlled by Approved Document (assume that minimum standard specified for purposes of SAP calculation)	Space and hot water system - TRVs, timer device and programmer on boiler, hot water cylinder with thermostat and timer, insulation to cylinder and pipework	4.41

Note: 'U' values for elements calculated using the 'Proportional areas' method and Tables A2, A4, A10 and C1 of Appendix A and Section 1 of Approved Document L.

The dwellings SAP rating has been calculated for designs under the requirements for both the 1991 and 1994 Regulations. The very low rating of **29** for the 1991 design can easily be attributed to the poor insulation levels of the existing fabric, high air infiltration rates, high ratio of single glazing to floor area and lack of control over the space and water heating. The improved rating of **55** under the 1994 specification is still quite inefficient and does not fulfil the requirements of Table 4 'SAP Ratings to demonstrate compliance' in Approved Document L. For a dwelling of 51m^2 floor area, it should be **80**. Paragraph 1.19 states: *'The requirement will be met if the SAP Energy Rating for the dwelling (or each dwelling in a block of flats or converted building) is not less than the appropriate figure shown in figure 4'.*

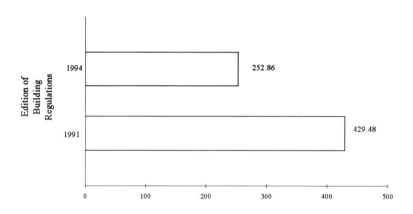

Figure 2 Effect of changes to Regulations on energy costs for dwelling

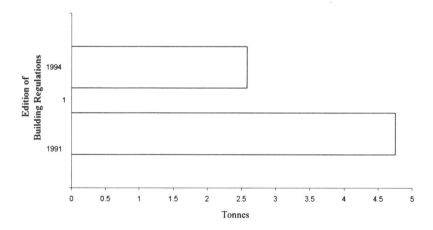

Figure 3 Effect of changes to Regulations on the CO_2 emissions of the dwelling

The fact that the existing windows were unable to be altered from single-glazed sashes due to controls under the Town and Country Planning legislation is a major contributor to the dwelling's inefficiency. These windows, together with the new openings, total 21.60m², **some 42 % of the floor area**. The 'basic allowance' for windows, doors and rooflights in paragraph 1.4 is for them to have an average 'U' value of 3.3 W/m²K and for their area to not exceed **22.5%** of the total floor area. For the dwelling in this example to remain with 42% of glazed areas, Table 3 of the Approved Document requires the 'U' value to be improved to less than 2.0 W/m^{2}K which is obviously expensive to achieve even if window alterations were permitted.

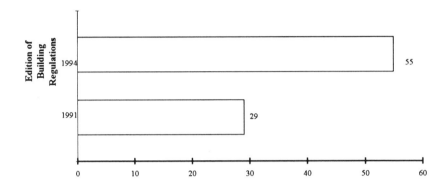

Figure 4 Effect of changes to Regulations on the SAP energy rating of the dwelling

This matter is a potential area of dispute in the requirements. On one hand the Document is asking for 'appropriate' figures for SAP ratings and on the other, it asks for upgrading work *of 'the extent ... (depending) upon the circumstances in each case'*. However, as new Regulation 14A requires all new dwellings to have calculated SAP ratings, it is possible that *market conditions* will be the determining factor. When energy rating becomes more widely understood by the general public, dwellings developed through changing the use of existing buildings may need to show equivalent levels of energy efficiency to compete in the marketplace. This, again, remains to be seen.

In the design of this particular dwelling, the internal volume could have been used more effectively to produce a two storey dwelling (and so reduce the window:floor area ratio) and/or to reduce the total area of glazing by specifying a solid door. These improvements would have raised the dwelling's SAP rating to **76**. To improve the energy rating even further, draughtstripping of existing windows and specification of a high efficiency boiler could have been incorporated raising the rating even further to **83** (this rating would be similar to that of a new build dwelling). The conversion work carried out in 1993 made an over-simplistic attempt to limit the building's energy use. If the designer had taken an holistic approach instead of focusing on specific aspects of energy conservation this may have been realised. Figures 2, 3, and 4 indicate the improvements to the energy efficiency of the dwelling. The *minimum* additional work required under the 1994 Regulations would have increased the SAP rating of the dwelling by 26 points, the carbon dioxide emissions for the dwelling are reduced by 54% per annum and energy cost savings of £176.62 are achieved. If the internal space of the original hospital building had been used more effectively and the windows and heating services improved beyond the requirements of the 1994 Regulations, the dwelling could compete in the same SAP rating levels as a new building.

The ability of the construction industry to meet the new standards

As stated earlier, the success of the new Regulations will depend upon the level of environmental education within the construction industry and the quality control and workmanship on site. When Danish Building Regulations were substantially improved, the government introduced a comprehensive training programme in the construction industry to raise the level of awareness concerning energy conservation technologies and building techniques. Unfortunately there was no such programme developed for the introduction of our new energy standards. Their announcement was made in sufficient time before their introduction to allow for training, however, this was carried out on a fragmented basis and mainly focused on training professionals to use SAP computer software. This is only a fraction of the environmental education which is required if the Regulations are to produce good quality, energy efficient buildings that will perform well over the coming decades.

On-site training is also required to achieve an adequate standard of workmanship. The Construction Industry Training Board have been contacted for their view on this matter. They comment that *'Our development staff responsible for training systems for technicians, supervisors and managers have developed training material based around BS7750. To date* (August 1995) *we have not found that the industry has been very receptive to that message.'*

If designers are not trained on a widespread basis and contractors are slow on the uptake of training in energy conservation measures, then surely it must be the Building Control Authorities who will ensure that adequate standards are met. However, there is already concern amongst building surveyors working within the field of building control that they may not have sufficient skills to implement the more advanced requirements in Approved Document L. Their education, historically, has not focused on energy conservation to any great depth. In addition to this, in the previous decade, they have had more pressing issues to deal with when the entire system of building control was changed enormously following the implementation of The Building Act 1984 and several other statutory duties have been altered and extended to increase their scope and depth of responsibility. Both of these matters have led to increased workload. This is further exasperated when the squeeze on Local Authority finances is taken into account. The drive to satisfy budget requirements has led in some cases to staff cuts and reductions in staff education and training for continual professional development (CPD).

It could therefore be argued that, in the drive to maintain an adequate level of building control and use their resources to maximum benefit, Building Control Authorities will prioritise their control to those requirements of the Approved Documents which endanger the health and safety of building occupants rather than the conservation of energy. This is not an unrealistic observation to make. Once a building is completed, defects in construction that affect the structure or fire safety aspects are much easier to detect than those hidden elements such as draught sealing, types of glazing, cavity insulation and cold bridging. Taking this theory one step further, it is also much less likely that work which falls below the standards for adequate energy conservation will lead to litigation against the Authority whereas failure to protect occupants of buildings from danger to their health and safety will.

Research from a variety of sources has shown that in many cases the 'as built' performance of buildings fall below the standards indicating poor design and/or poor workmanship. At present there is no *statutory* provision in the Regulations for inspection of energy conservation measures during construction or random post-completion checking of buildings such as pressure testing for air leakage and thermographic surveys to check heat transfer rates through the building fabric.

The question must therefore be asked, *is the government introducing energy conservation standards which designers, contractors and building control surveyors are incapable of achieving?* The answer to this should be NO.

CONCLUSIONS

Upgrading the requirements of the Building Regulations is only one of a package of measures implemented by government to curb the UK's emissions of CO_2 from energy used in buildings. This paper has examined the effect that extending the control of Regulation 6 (1)(a) to cover energy conservation for change of use work will have on energy savings in the existing building stock. The percentage of such work, at 4% per year, indicates a very small sector of construction work. However, the case study has shown that substantial energy savings can be achieved through changing the of use of old, inefficient buildings with relatively little technical risk.

New build work accounts for a minute proportion of the overall building stock. If significant progress is to be made in promoting energy efficiency it is essential that existing buildings are tackled.

The Regulations which came into effect on 1 July 1995 have three main advantages over the previous Regulations:
1. the scope of the regulations has been broadened,
2. the standards for heat loss have been tightened, and

3. a standard assessment procedure has been introduced which provides a quantified indication of the efficiency of new dwellings.

As the SAP energy rating system becomes more widespread, it is reasonable to assume that the owners and occupiers of dwellings will become more aware of the efficiency of property. This increase in information should have the benefit of providing an economic incentive for buyers to choose upgraded property. This market led incentive may become more powerful than the Building Regulations in promoting energy conservation for *converted* property as it has been shown that the minimum requirements in Approved Document L do not produce dwellings with equivalent SAP ratings to those for new build.

The success of the new requirements relies on the level of environmental education in construction professionals and the workmanship and quality control on site. It is regrettable that the scope of the Regulations does not cover all alteration work and that the Approved Document L focuses only on the conservation of fuel and power in heating and lighting installations. It is interesting to note that the control over mechanical ventilation and air-conditioning has not been included in the revisions, therefore the original emissions target provided for Building Regulations, at 0.25 million tonnes of carbon, should also be reduced to take this into account.

Designs which are required to alter or convert a building are inherently more difficult than new build developments due to the restrictions of the existing building. When considering energy conservation under these circumstances, an audit of the building is essential to determine the features which will work for energy conservation and those that will work against it. The surveyor carrying out the initial inspection of a building proposed for alteration must therefore be trained to audit the building's energy features. The person responsible for the design and specification should be educated in the most suitable methods for improving energy efficiency. Simply designing to the minimum standards set out in Approved Document L may not be appropriate or adequate for existing buildings. A holistic approach is required which may involve more thorough calculations and a more imaginative design approach. Once a design is completed it is also essential that the level of workmanship and quality control on site is capable of ensuring the finished building will perform as expected.

RECOMMENDATIONS

Now that we have a reasonable standard set out in the Approved Document L it is essential that the Building Regulations themselves and the people responsible for their implementation are equally prepared. This paper has pointed to a number of factors which may prevent the full benefit of the improved standards being realised. It is therefore recommended that further research be carried out into:

- how *Regulations 3(2)(a) and 5* can be amended to cover a wider range of alteration and conversion work (especially in the case of office buildings), and

- how *Regulation 2* can be changed to allow lighting installations to be defined as a 'controlled service' for material alteration work.

- *Regulation 14* 'Notice of commencement and completion of certain stages of work' and *Regulation 15* 'Completion Certificates' could be extended to include a thermographic survey and air pressure tests of property, particularly housing, to establish points of cold bridging and air leakage.

Finally, the recommendation of the House of Commons Select Committee on the Environment to promote better training and education for professionals should also be given much more consideration than it has been. The RICS can play an important role in this respect.

REFERENCES

BRECSU (1994). *The Government's Standard Assessment Procedure for Energy Rating of Dwellings.* 3rd edition. BRE,Watford.

Centre for Configural Studies (1994). *Non-domestic building stock project report.* Open University, Milton Keynes.

Department of the Environment (1991*). English House Condition Survey, (1986) - Supplementary (Energy) Report.* HMSO, London.

Department of the Environment (1992). *Good energy practice guide 33 - energy efficiency in offices.* HMSO, London.

Department of the Environment (1994a), *Climate Change - The UK Programme.* HMSO (CM 2427), London.

Department of the Environment (1994b). *The Government's Response to the Fourth Report from the House of Commons Select Committee on the Environment - 'Energy Efficiency in Buildings'.* HMSO (CM2453). London.

Department of the Environment, (1994c*). Building Regulations, Approved Document L: Conservation of Fuel and Power.* HMSO, London.

Department of the Environment Consultative Committee on Construction Industry Statistics (1995). *The State of the Construction Industry.* DoE, London.

Department of Trade & Industry (1994). *The Energy Report Volume 1 - Markets in Transition.* HMSO, London.

Department of Trade & Industry (1995). *Energy Paper 65, Energy Projections for the UK - Energy Use and Energy - Related Emissions of Carbon Dioxide in the UK.* HMSO, London.

Energy Efficiency Office (1992). *Good Practice Guide 33 - Energy Efficiency in Offices.* HMSO, London.

House of Commons Environment Committee (1993). *Session 1992-93 - Energy Efficiency in Buildings.* HMSO, London.

National Audit Office (1994). *Buildings and the Environment.* HMSO, London.

RICS, (1990). *Building Surveying Workload Statistics Survey.* RICS, London.

Spring, Martin (1994), *Leaner & Fitter, Building Renewal,* 29 September, pp6-7

7 The Valuation of Contaminated Land and Property

TIM DIXON & TIM RICHARDS
College of Estate Management, Reading, UK

INTRODUCTION

Contamination of land is problematic because it presents risks to the health of living organisms and also creates financial difficulties and uncertainty for businesses. Substantial areas of the UK are currently contaminated and recently, sizeable disposals of such land have been made by British Coal and British Gas (Jenkins, 1995). Given this, and in the wake of the Environment Act 1995 and the mounting 'green concerns' of many large businesses, it is clear that the issue of contamination is one that is of great significance to the valuer. This paper seeks to outline a set of 'best practice' approaches which valuers can use in the valuation of contaminated land and property.

The paper will focus upon the valuation and the calculation of worth of both fully-let and reversionary freehold investment properties. A distinction is thereby drawn between valuation (the estimation of open market value or the most likely selling price) and the calculation of worth (the estimation of net monetary worth to an individual, of the benefits and costs of ownership) (RICS, 1995a). In particular, the paper will investigate the suitability of the 'cost to correct' approach to the valuation of contaminated land, and the roles which the comparative, investment and discounted cashflow (DCF) methods play within this. Although the paper focuses upon fully-let and reversionary freehold investment properties, it is anticipated that the general principles which are discussed in relation to valuing these properties can be applied, with modification, to other forms of contaminated property.

Following a background section, the methods used for the research and selective results will be presented and analysed. Further details of the full results can be obtained from 'A Changing Landscape: The Valuation of Contaminated Land and Property' (Richards (1995)).

BACKGROUND

This section aims to provide an understanding of the main terms, legislation and research and guidance which are considered to be important to this area of study.

The Environmental Impact of Land and Property Management, Edited by Yvonne Rydin
© 1996, The Royal Institution of Chartered Surveyors

DEFINITION AND EXTENT OF THE CONTAMINATION PROBLEM

A definition

Arriving at an agreed definition for contaminated land is fraught with difficulty. Tromans and Turrall-Clarke (1994) noted that the distinction which can be drawn between 'contamination' and 'pollution' is often ignored. ('Contamination' can be caused by the mere presence of a foreign substance, whereas 'pollution' results when such a substance causes harm.) Another aspect of the definition which has been debated, is whether the test for contamination should be based upon the use to which land is to be put, or on the harm which substances within the land may present to man or the environment.

For the purposes of this paper, contaminated land will be defined in accordance with the Environment Act 1995, as being:

> 'in such a condition, by reason of substances in, or under the land that -
> (a) significant harm is being caused or there is a significant possibility of such harm being caused; or
> (b) pollution of controlled waters is being, or is likely to be caused...'

In terms of 'harm', the main concerns are:

> '...the health of living organisms or interference with the ecological systems of which they form part and, in the case of man, includes harm to his property...'

Extent of the problem

The extent to which contamination presents a problem in the UK is uncertain. The House of Commons Environment Committee published its first report on contaminated land in 1990, and estimated that there were between 50 000 and 100 000 potentially contaminated sites in the UK. The CBI (1994) claimed that there were some 200 000 hectares of contaminated land in the UK. However, these estimations depend greatly upon the assumptions made about the contaminative potential of each use, the risks posed to health by contamination, or the extent of contamination at each site.

Problems for valuers

The issue of contamination has presented valuers with a number of problems. For example, the profession has not yet fully come to terms with the comparatively new issue of contamination, because little is known as yet as to what impact it holds for property value. In this sense, there is not only a shortage of comparable information, caused partly by the property recession, but there is also uncertainty. This uncertainty has resulted, in part, because comprehensive legislation has been more recently implemented than, for example, in the USA (Stephenson, 1993). MacRae (1994) illustrated the uncertainty of valuers, when he reported that funds were 'too negative' about polluted land. He therefore maintained that the message needed to be conveyed to them that contamination is manageable, and that decontaminated sites should be valued accordingly.

There is therefore uncertainty as to which valuation methods to apply in the valuation of contaminated land and how these methods should be adapted for use in often unique circumstances.

One of the main problems is the fact that, in most instances, valuers do not have the necessary skills, nor indeed the professional indemnity insurance cover to enable them to assess the environmental conditions of the ground and the environmental risks attached to properties and sites.

It may appear that in attempting to resolve such issues, too much responsibility is being placed on the valuer. However, it must be remembered that the remit of valuers should be to offer comprehensive services in relation to all types of land and property. Furthermore, the value of land and property is of critical importance to many parties - and it is therefore in everyone's interest to ensure that the price paid for it accurately reflects its true worth (National Westminster Bank, 1994). If valuers too easily decline instructions where contamination is considered to be an issue, then they may lose out to other professions offering alternative advice.

Legislation

Legislation is important in the context of contaminated land valuation, because present and prior owners and lenders may be adversely affected by environmental liability. For example, law may dictate that land or property may need to be cleaned up by the current owner to a particular standard and transfers of land may create a liability for a future owner or occupier. Such liability regimes in relation to contamination are therefore likely to reduce the value of the property.

A key statute in the area of contaminated land is the Environment Act 1995. This Act was largely based upon approaches introduced in the DoE consultation paper, 'Paying for our Past' and the resultant 'Framework for Contaminated Land' (DoE, 1994a,b). These documents were, in turn, the production of an intensive interdepartmental review, which was set up after the government decided to abolish the proposed s.143 registers. These registers were to have detailed any land subject to past or present contaminative uses. However, proposals were formally repealed with the introduction of the Environment Act, following criticisms from landowners, property developers and financial institutions who pointed to the potential blighting effect and resultant devaluation of their assets.

One of the most relevant aspects of the new Environment Act 1995, in terms of its impact upon the land and property industry, concerns the duty which it places upon local authorities (LAs) to inspect land within their areas and seek out those areas of contaminated land (in accordance with the definition presented earlier). The enforcing authority (either the LA or the newly formed Environment Agency) must then serve remediation notices upon the 'appropriate' persons, in relation to any areas of contaminated land.

The 'appropriate person' follows the 'Polluter Pays Principle', and will be the person or persons who caused or 'knowingly permitted' the contamination to occur. If such a person cannot be found after 'reasonable enquiry', the appropriate person will then be the 'owner or occupier for the time being of that contaminated land in question' (Barrett, 1995). 'Owner' is defined within the Act in terms of entitlement to receive the rack rent (essentially, the market rent yielded by the land if let), and explicitly excludes any mortgagee who is not in possession of the land. 'Occupier' is undefined in the Act.

Records will also be kept of those properties upon which remediation notices have been served. In this respect, 'registers' of contaminated land are to be kept, but these will instead be of land which is actually contaminated, rather than of any land subject to potentially contaminative uses.

Professional guidance

The RICS has produced two guidance documents relating to contamination, and valuers need to have a good knowledge of these documents to operate efficiently.

The first, Valuation Guidance Note 11 (VGN11), entitled 'Environmental Factors, Contamination and Valuation' (RICS, 1993), confirms that, unless otherwise agreed in advance, the valuer should reflect all relevant issues (including contamination) within a valuation, and should not automatically assume that a property is uncontaminated. Guidance is then provided as to the type of enquiry and sources of information which valuers can use to assist them in valuing contaminated properties.

VGN11 details three caveats which may be included within valuation reports of this type:

Caveat 1   for instructions which are to disregard contamination.
Caveat 2   for instructions where enquiry is made but there is no evidence of
              contamination; and
Caveat 3   for instructions where there is evidence of contamination, and the
              remediation cost has been estimated.

At the time of writing, a new RICS Appraisal and Valuation Manual has been produced (RICS (1995a)). This manual combines and replaces the previous red and white books, and becomes a mandatory document on 1 January 1996. Within this new 'Red Book', VGN11 has been replaced by a series of new documents, the details of which are discussed in the conclusions of this paper.

The second guidance document, 'Land Contamination Guidance for Chartered Surveyors' (RICS, 1995b), provides surveyors with a useful overview as to how they should deal with environmental issues, including what to look for and how to provide advice relating to contaminated land. The idea of a Land Quality Statement (LQS), which is the written output of an environmental risk assessment, is also introduced. The guidance document stresses how the LQS can be used as a tool for the assessment and management of environmental risk.

THEORETICAL RESEARCH FROM THE USA

There is a substantial body of research literature relating to contaminated land and valuation. Perhaps the most developed in this respect is from the USA, which experienced the earlier introduction of more rigorous and potentially onerous environmental legislation. There is thus a great deal which can be learned from the other side of the Atlantic (Dixon, 1995 and Dixon and Richards, 1995).

A significant number of theoretical articles covering the valuation of contaminated property have appeared in the US *Appraisal Journal*. A seminal article in this respect is that by Chalmers and Roehr (1993), in which the authors indicate that the effect of contamination upon property value must be examined within a systematic framework incorporating such factors as the:

- nature and extent of contamination
- way in which the problem is perceived or evaluated
- remediation and indemnification responses to the contamination
- the effect of these responses on utility and marketability, and
- the value consequences.

In terms of actual valuation methods which may be employed for the valuation of contaminated land and property, Wilson et al. (1993) examined the applicability of the 'sales comparison approach' and Patchin (1988 and 1991) and Mundy (1992a, b and c) developed models to take account of contamination using an income approach.

At the heart of their work is the 'cost to correct' (or 'cost to cure' approach), which takes an uncontaminated value figure, using either the investment method or comparative method, and deducts remediation costs. The end result is the contaminated value. A simple example of this is shown below:

Income	£100,000
YP Perp @ 10%	10
	£1,000,000
Less:	
Remediation Costs	£300,000
Value	**£700,000**

Such a simplistic approach can, however, overlook the phenomenon of 'stigma'. Wilson (1994), Patchin (1988, 1991) and Mundy (1992a,b,c), however, address this issue of 'stigma' and its importance in the valuation of contaminated land.

In terms of a definition, Wilson (1994b), succinctly describes 'stigma' as the 'value impact of environmentally-related 'uncertainties' and Patchin (1991) describes it as a 'negative intangible', caused by fear of hidden clean-up costs, the 'trouble' factor associated with work involved in clean-up, the fear of public liability, and the lack of mortgageability.

It is suggested that this 'stigma' element can be incorporated into valuations by making an upwards adjustment to the yield figure adopted or by making an end deduction or allowance to the valuation or calculation of worth. This could be incorporated into the previous valuation example as shown below:

Income	£100,000
YP Perp @ 11%	9.091
	£909,091
Less:	
Remediation Costs	£300,000
Value	**£609,091**

Such an adjustment clearly reduces the capital value figure. (In this case, a 1% increase in the yield to reflect 'stigma' has created a 13% decrease in capital value.) The derivation of such a mark-up, however, is fraught with difficulty. In the absence of comparable evidence, subjective judgement can play a major role.

THEORETICAL RESEARCH FROM THE UK

There has been relatively little research into contaminated land valuation in this country, in comparison to that which has been conducted in the USA.

Sheard (1992 and 1993) is one of the few to tackle more theoretical issues in the academic and property press. He observes (1993) that:

> '...Valuing contaminated land is a fairly basic application of traditional valuation theory ... there is, however, no ... accepted approach to valuing contaminated sites...'

He argues (1992) that a 'mini-residual' or a 'before and after approach' (allied to the 'cost to correct' approach) may be used. However, this, as he later acknowledges in Tromans and Turrall-Clarke (1994:587), will take no account of 'stigma'.

More recently, Lizieri et al. (1995) have pointed out that in England and Wales, although the problem of contamination has been recognised, valuers have been less able to propose solutions. In their view, the use of an all-risks-yield approach holds its own perils of 'double counting' in the adjustment of yield for specific environmental risk, where the asset class should already affect the risk.

Several surveys of UK professional practice have also been conducted. These include Environmental Assessment Group (EAG) (1993), Hillier Parker (1993 and 1994) and Lizieri et al (1995).

These surveys indicate that although the property profession is slowly coming to terms with contamination issues, a great deal still needs to be done in order to create more certainty and therefore consensus in terms of valuation approach.

SUMMARY

A synthesis of the results of research literature on both sides of the Atlantic suggests that there is a logical or systematic approach to the valuation of contaminated freehold investment properties, which valuers theoretically should adopt.

In particular, in relation to freehold investments, valuers will need to utilise a 'cost to correct' approach, having first utilised the comparative and investment methods. A key issue in this, is the presence of 'stigma'. Valuers may decide to ignore this or to account for it explicitly. Exactly how they do so in practice for this and other costs was investigated in the empirical research conducted for this paper, which is described in the next section.

THE RESEARCH STUDY

Aims of the research

The specific aim of the research which this paper describes is to formulate a set of 'best practice' approaches which valuers can use to guide them when valuing contaminated (or potentially contaminated) investment properties.

'Best practice', in this context, is defined as 'best market practice' - i.e. those valuation methods and procedures which could most realistically and beneficially be applied by valuers in the market when valuing contaminated land or property.

The key objectives incorporated within this overall aim are to analyse the theory and methodology of different methods of valuation, as applied to contaminated land and property and to investigate the issues of 'stigma' and 'remediation costs'.

The paper will focus on the valuation of fully-let and reversionary freehold investment properties, although it is anticipated that the general systematic approaches adopted can also be applied, with modification, to other forms of property which may be contaminated.

Research methods used

Two main research methods were employed in producing this paper.

Firstly, 50 structured face-to-face interviews were undertaken with a wide variety of professionals, from those sectors which are active in dealing with contaminated land and property. The interviews were conducted across England and Wales, to give the paper a representative national focus. As this paper aims to examine valuation specifics, only the 30 interviews which were undertaken with valuers will form the focus of discussion. This sample was also chosen so that valuers from small, medium and large firms were interviewed. The individuals selected for interview were invariably the senior partners or directors of companies, or those with the greatest experience in relation to contamination issues. The questionnaire for valuers contained 35 questions which covered a number of key areas.

This paper will investigate those areas which are considered central to valuation, including the main factors, procedures and methods in relation to contaminated land valuation, together with those factors which should be considered prior to any valuation.

Secondly, a more detailed analysis of valuation specifics was also enabled through the formulation of a number of hypothetical valuation scenario questions. These questions were devised so as to incorporate contamination issues and were presented as pro-formas and given to those valuers who were interviewed. It was intended that the questions would reveal valuation practicalities, including the approaches used, methodologies employed and actual adjustments made by valuers in relation to a fixed set of valuation problems. These responses were then analysed and compared in order to produce a set of best practice responses to the valuation scenario questions. It was further anticipated that this more quantitative approach would support and add focus to the information from the more qualitative, structured interviews. (The actual questions which were asked are presented in full in the results sections.)

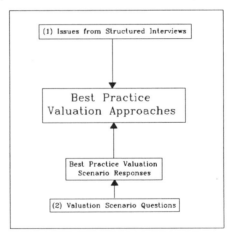

Figure 1 Research design

Figure 1 shows how these two methods will be combined to produce best practice valuation approaches. The results from these two methods will, therefore, also be presented in this format.

RESULTS FROM STRUCTURED INTERVIEWS

This section presents the results which were obtained from the structured interview questionnaires and will focus on a discussion of:

The issues relating to contamination which the valuer should consider as a pre-requisite to valuation.

The main 'cost factors' which the valuer will need to calculate in order to produce a valuation of contaminated property.

The relative appropriateness and adaptability of different valuation methods into which the 'cost factors' can be input, in order to produce valuations or calculations of the worth of contaminated property.

Actual quotations taken from the interviews are used to illustrate specific points.

Factors as a pre-requisite to valuation

Before a valuation of contaminated land or property can be produced, the environmental risk relating to that land or property must be assessed.

It is vital that the valuer checks the limitations of his Professional Indemnity (PI) insurance cover, as this may not enable him to produce such an assessment himself.

If the valuer does not possess adequate PI cover, then he should advise the client that an environmental consultant should be employed, in order to assess this risk element. The results from this assessment can then be used as an input into the valuation process. In this way, a 'team approach' can be adopted in the valuation of contaminated land and property.

Even if he does not have the necessary expertise or PI cover, it is important that the valuer has an understanding of what environmental risk assessment entails. Such an understanding will enable him to adequately brief an environmental consultant on behalf of the client, thereby facilitating this 'team approach' and also objectively interpret any results and incorporate them into his valuation or calculation of worth. Perhaps most importantly, a better understanding of this process should also enable the valuer to more easily recognise when contamination may present a problem in relation to a property and consequently, when he should seek more specialist advice. An environmental risk assessment can take the form of an LQS as described earlier .

The environmental risk relating to a property can be seen as a function of the nature, degree and extent of any contamination, viewed in the context of the end use of the site. In this sense, given similar contamination, the standard of remediation which would be required for a car park would not be as stringent as that required for a residential development.

One of the interviewees gave an appropriate summary of this:

'It is a question of contamination levels being within acceptable levels - you can live with contamination in relation to many uses.'

A further element to risk assessment is the application of the 'Substance - Pathway - Target' model, as introduced in Land Contamination Guidance for Chartered Surveyors (RICS,1995b). An environmental risk assessment (or LQS) should thereby determine any potential sources of contamination (both on the property in question and neighbouring properties) and those substances which may emanate from these sources. In addition, pathways by which contamination may migrate to, from or around the site (such as water courses) and any targets which may thereby be affected should be considered.

Such an approach to the assessment of environmental risk should ensure more holistic and realistic end results.

Main 'cost factors'

The main cost factors which the valuer will need to calculate when valuing contaminated property were found to fall into three groups: direct costs, indirect costs and property factors.

Direct costs

In terms of direct costs, the main issue is the cost of remediation and monitoring. It is also important to include within this figure, where appropriate, any investigation or consultation fees which may have been necessary to determine the need for remediation works and monitoring.

Depending upon the magnitude and extent of the contamination problem, and the response of the owner to the problem, it may also be necessary to incorporate under this heading the costs of financing remediation works, 'void' costs (where occupied buildings may need to be vacated to enable remediation: in a let property this will mean the incorporation of a rental void period within the valuation); and any penalties for non-compliance with legal requirements (such as a remediation notice).

Indirect costs

Indirect costs incorporate any 'stigma' risks or contingencies which may be made in relation to a property.

This is the most subjective and difficult element to determine and cost, but, nonetheless, it may have a significant impact upon valuation, and may remain with a property even after remediation. One interviewee noted that he had seen discounts of 25% to 30% made on uncontaminated capital values because of this factor.

The importance of uncertainty in relation to stigma in property valuation, is emphasised by the quotations which are presented below:

'It's the uncertainty factor: risk which is difficult to quantify. I mean you can quantify the [remediation] costs.'

'It's easy to say that the [value] impact of contamination is simply the cost of taking it away - but there's no such thing as taking it away - you can never take it away.'

'The chances are that if there is serious contamination, the investor simply won't buy because there are too many imponderables, they can't value it, they can't assess the risk, they want a hassle-free investment - so effectively, it tends to kill it stone dead, as HAC [High Alumina Cement] used to!'

Property factors

A final important, and often over-riding, group which should be considered in this sense is 'property factors'. These factors include the effects which location, supply and demand may have upon price.

The old adage of 'location, location, location' being one of the most significant influences upon property value also seems to apply in relation to contaminated properties. If the property is in a desirable location, then the issue of contamination may be viewed less critically.

Similarly, if there is considerable demand for a property, then buyers may feel compelled to work with it, in spite of it having contamination problems.

'There are many factors in property which will override certain things in certain situations - if someone wants something [a contaminated property] badly enough, they will take a risk on it.'

It is therefore vital that these 'property factors' are considered in any valuation calculation, due to the significant effects which they may have upon value.

Finally, it must be stressed that contamination is only one of many aspects which may go towards forming a valuation, and, in practice, it may be difficult to disaggregate the value impact of the different factors.

'There isn't a straight academic answer - it is one of the ingredients that goes into the mixing bowl. It [contamination] probably isn't a killer on its own, but there are a million and three ingredients that go into any acquisition. There is no one thing that kills a deal, but it is often a combination of things.'

Methods for the valuation and the calculation of worth of contaminated properties

This section will examine the suitability of the 'cost to correct' approach (presented earlier), citing the example of the valuation of contaminated fully-let freehold properties, and in the context of views expressed by interviewees.

The most suitable methods to use in producing calculations of worth, as opposed to valuations, of such properties will also be examined.

Valuation methods

The 'cost to correct' approach relies upon obtaining comparable income and yield figures. The valuer must decide whether to use comparable information from impaired (contaminated) or unimpaired properties. At this stage, stigma may or may not be reflected in the all-risks-yield that is used (see below). A distinct danger in using impaired information is that it is highly unlikely that two contaminated sites will be comparable:

> *'It depends upon what the nature of the problem is. It isn't necessarily consistent over a whole area. If you take two greenfield sites - they are comparable, there is also plenty of evidence - but two brownfield sites - the nature and amount of contamination and the location of it could be so different that the comparison is impossible.'*

It is therefore more realistic to treat each site on its own merits, seek out unimpaired comparable information, and then make adjustments to the figures used.

Having said this, it may be more suitable to use impaired comparable information in those more industrialised areas, where contamination of land can be seen to be a more homogenous phenomena. However, this can also be a very dangerous assumption to make, as not all land is equally contaminated:

> *'I think you can start to use the comparative method, perhaps more in the Black Country where there is plenty of contaminated land...although you have to make the all-in-one global assumption that all the land in the area is equally badly contaminated.'*

The next stage within the 'cost to correct' approach, is the capitalization of an income figure, at an all-risks-yield (ARY), in order to produce a capital value figure.

It may be suitable, however, to reduce the income figure if there are any on-going liabilities which arise due to contamination. Such liabilities may include annual monitoring costs, as these would effectively create a reduction in the rent received.

The final stage of the 'cost to correct' approach is the deduction of remediation costs from the capital value figure obtained. Some argue that it is more suitable to make an explicit end deduction for remediation costs, than to make an adjustment to the ARY figure, because such adjustments are highly subjective:

> *'Yield adjustments can be too ad-hoc.'*

Such an approach, however, ignores stigma which may not be perceived as a problem in any case. However, if stigma is deemed to be important, then this can be reflected within the valuation by making an appropriate adjustment to the ARY figure adopted, or by making an end deduction or allowance to the valuation or calculation of worth.

How though can such an adjustment be made? A number of implicit assumptions may be contained within one yield figure (such as to growth, risk of illiquidity, and risk of tenant default). An adjustment to this figure therefore creates yet another implicit assumption and there is hence a corresponding difficulty in disaggregating these factors from one another. However, it may be possible to gather comparable information to assist in its calculation, especially as more contaminated and previously contaminated properties are traded in the market.

One respondent advocated this approach, saying;

'I would want to capitalise the rent at a yield that was higher (i.e. a lower YP) to reflect:

(a)   the stigma which would impact possibly on re-lettings

(b)   to reflect a potential future remediation cost (a contingency allowance if you like), but to allow for the fact that the remediation may break down and in time you would have to do it again, and

(c)   to reflect the loss of flexibility or rather the re-development planning being rather more difficult.'

A further point, which must be appreciated when making such adjustments to the ARY, however, is that an increase of, say, 1 percentage point will hold widely different value impacts for investment properties which have different uncontaminated ARYs (see example below).

Contamination can therefore hold different value impacts for different end uses. However, in the situation above, contamination would arguably present a similar risk because in both cases there is little chance of humans coming into contact with the soil. The greater percentage reduction in the value of the contaminated shop investment, as opposed to the industrial investment (27% as against 19% reduction from unimpaired value), may therefore be an over-representation (in financial terms) of the real risks which contamination presents.

For this reason, adjustments to account for stigma should be made as a proportion of the ARY figure. Applied to the example (overleaf), an equivalent adjustment to the yield for the retail property would result in an increase of +0.5%, which would give a reduction in value of 19%, commensurate with that of the industrial property.

Eliot-Jones (1995) noted after the Bixby Ranch Case, in California, USA, that similarly absurd results can be produced where stigma is calculated as a percentage of unimpaired value for an investment property.

Example - Two investment properties with non-migratory heavy metal contamination

(1) = An industrial property in the North of England
(2) = A retail property in central London

(A)   **Value ignoring contamination**

(1)	Income	£100,000	(2)	Income	£50,000
	YP Perp @ 10%	10		YP Perp @ 5%	20
	Value	£ 1,000,000		Value	£ 1,000,000

(B)   **Value including contamination**

(1)	Income	£100,000	(2)	Income	£50,000
	YP Perp @ 11%	9.09		YP Perp @ 6%	16.67
	Total	£909,091		Total	£833,333
	Less:			Less:	
	Remediation	£100,000		Remediation	£100,000
	Value	**£809,091**		Value	**£733,333**
	Reduction	**19%**		**Reduction**	**27%**

Methods for producing a calculation of worth

The suggestion of applying DCF methodologies met with mixed reactions amongst the interviewees. Some saw it as a very useful technique for analysing the worth of contaminated investments, as it enabled more realistic phasing of the many costs which may be incurred in relation to contaminated properties (such as for site investigations, remediation and monitoring). The technique was also considered to be very useful in that the internal rate of return figure produced can enable direct comparison with other alternative investments.

> *'DCF is going to be increasingly important. You are looking at very very large sums of money for decontamination so there are elements coming in, not just the time elements, but you have also got the taxation or tax allowable element of that, which is going to have to be factored.'*

> *'With DCF, you can compare an internal rate of return of various investments which would have a different all-risks-yield.'*

Other interviewees were less convinced as to the usefulness of the method:

'DCF is a waste of time - because you've got no way of forecasting future income and relatively little comparable evidence for discount rates.'

A final point to consider in terms of valuation methodology is, as one of the interviewees maintained, that the valuation method chosen should relate to the property type which is to be valued, rather than to whether or not contamination may be an issue.

'We don't tend to value a property or site and choose the method of valuation depending upon whether or not it is contaminated, because I feel that would be incorrect. We will choose a method of valuation that we feel is appropriate for that particular property irrespective of whether or not we feel it might be contaminated. If the site did happen to be contaminated we would then have to deal with how that impacts upon the value, if at all.'

VALUATION SCENARIO QUESTIONS

The valuation scenario questions were designed to highlight how different types of property with different types of contamination could be valued and also how the worth of properties could be approached. All three questions which will be discussed here concern the valuation of reversionary freehold investment properties:

Question (1) concerns an office investment affected by non-migratory heavy metal contamination.

Question (2) concerns an office investment affected by groundwater contamination.

Question (3) asks for a calculation of worth of an office investment affected by non-migratory heavy metal contamination.

A total of 12 valuation scenario responses were received out of the 30 which were sent to those valuers which were interviewed: this represented a 30% response rate. The responses indicate some of the practical issues concerning the ways in which valuers approach the valuation of contaminated property, and also indicated the magnitude of some of the adjustments that were actually made.

The questions will be approached sequentially. Firstly, the responses which were received will be summarised.Secondly the best practice responses which were defined, will be presented.

These best practice answers were deduced following a discussion and synthesis (by an 'expert' panel of valuers assembled to advise the research team) of those responses which were received.

Question (1)

You are acting for a client who wants to know the price which he should pay for a freehold office investment which lies upon soil which is contaminated with non-migratory heavy metals.

You are given the following information:

* *The party responsible for the original contamination is the current freeholder (the developer) and the tenant is not liable at all.*
* *The current rent passing on the proeprty is £80 000 pa, with 2 years unexpired on a 25 year lease (5 year reviews). Remediation is required at the end of the term, and it is assumed that the present tenant will not renew.*
* *Present-day remedeation costs of the site to a 'suitable for use' level have been estimated (following an 'intrusive' environmental investigation) at £200 000.*
* *All-risk-yields and full rental values on similar uncontaminated properties currently stand at 10% and £85 000 pa respectively.*

Results

All the respondents applied the investment method in a term and reversion format when answering this question and then deducted remediation costs from the capital value figure that was obtained (as in the 'cost to correct' method).

The majority of respondents (92%) also allowed for a void period within their calculations, so as to enable the actual remediation work to be undertaken and to allow for re-letting, and finding a new tenant upon lease renewal. Of those respondents who included a void period within their calculations, two-thirds opted to make this a one year void, whilst the other third made this a two year void. The void periods were integrated into the calculations by discounting the reversionary income for 3 or 4 years (depending upon whether a 1 or 2 year void period was chosen).

83% of the respondents also reflected the additional perceived risk or 'stigma' in relation to contamination within their valuations. In the majority of these cases (75%), this was done by making an upwards adjustment to the yield figures that were adopted for the term and reversion parts of the valuation. There was no exact consensus amongst the responses as to what this increase should be, nor indeed which, vis-à-vis the term and reversion parts of the valuation, should have the greater yield increase. All the yield figures which were used, however, fell within the range 10% to 12%.

The other respondents (25%) allowed for this 'stigma' or uncertainty element by making end deductions or making contingencies on the remediation costings which were given.

Best practice answer

The best practice answer that was devised for question (1) together with a brief summary is presented below.

Firstly, the yield figures which were used for the term and reversion elements of the valuation were increased from the uncontaminated yield of 10%, by 1% to reflect the uncertainty which contamination may present at re-letting and by an additional 0.5% to reflect a post clean-up stigma. This gave a total 'contaminated' yield of 11.5%[2]. However, in keeping with the custom of the term and reversion method, the yield for the term element was then reduced by 1%, so as to reflect the greater relative security of the term income [1].

Qu. 1 - Non-migratory heavy metals

Term Income	£80 000	
YP 2 yrs @ 10.5%[1]	1.724	**£137 920**
Revisionary Income	£85 000	
YP Perp @ 11.5%[2]	8.696	
PV £13 years @ 11.5%	0.721	**£532 934**
Total:		**£670,854**
Less:		
Clean-up costs	£200 000[3]	
Finance on Above		
(6 months @ 10%)	£ 10 000	
Intrusive investigation	£ 20 000	
Cost of LQS	£ 2 500[4]	
Total	£232 500	
PV £1 2yrs @ 11,5%	0.804	£186 930
Value:		£483 924
	Say	**£484 000**

A void period of one year was also integrated into the calculation, by discounting back the reversionary income, as had been done in the majority of the responses.

A number of additional costs were also included within the best practice calculation, which none of the respondents had included:

Firstly, finance costs were included on the clean up costs on an 'S-curve' basis. This was at a rate of 10% for 6 months [3].

Secondly, the costs of an intrusive investigation and for the preparation of a Land Quality Statement were included within the calculation [4]. This was because it was considered that a purchaser would want to check the degree and extent of any contamination prior to purchasing such a property. It would also, following investigation, be prudent to prepare an LQS to generate a greater degree of certainty in relation to the value of the property.

Question (2)

*Assuming there is groundwater contamination, and that all other factors are the
same as Question (1) (above), how would you take account of this in your valuations
(and advice)?*
    Please explain how you would quantify the difference in impact (if any).

Best practice answer

Qu. 2 - Groundwater pollution			
Term Income	£80 000		
YP 2 Years @ 11.5%[1]		1.701	**£136 080**
Reversionary Income	£85 000		
Less monitoring costs	£ 2 000[2]		
	£83 000		
YP Perp @ 12.5% [3]		8	
PV £1 3Years @ 12.5%		0.702	**£466 128**
Total:			**£602 208**
Less:			
Clean-up costs	£200 000[4]		
Finance (6 months @ 10%)	£ 10 000		
Intrusive investigation	£ 20 000		
Cost of LQS	£ 2 500		
Total	£232 500		
PV £1 2yrs @ 12.5%		0.790	£183 675
Value:			£418 533
		Say	£419 000

The best practice answer is similar in format to that presented for Question (1).

It was felt (as indicated by respondents and the expert panel), however, that changes
should be made to this valuation in order to reflect the valuation impact of a more
serious contamination problem.

Firstly, it would be necessary to carry out monitoring of the site, as groundwater
contamination is highly migratory. Monitoring would therefore track the success of

any remediation works. Monitoring costs would be borne by the purchaser on an annual basis and these were therefore deducted from the reversionary income figure [2].

Secondly, the yield figures that were used in the calculation were increased by a greater amount than was the case in Question (1), so as to reflect the greater perceived risk. Therefore, the residual stigma element was increased from 0.5% to 1.5%, which together with the 1% premium (added for uncertainty regarding re-letting) produced a 'contaminated' yield of 12.5% [3]. As with question (1), this yield figure was reduced by 1% in the term, in order to reflect the greater security of income [1].

Results

Although the overall format of this question remained the same, all respondents recognised that this type of contamination presents a significantly greater risk. Indeed, many of the respondents indicated that it could be impossible to delineate such a problem, as groundwater contamination may migrate to other sites, and several also stated that they would fear future action by the Environment Agency. Therefore, 75% of the respondents maintained that groundwater contamination could present an unacceptable risk in this instance and that they would either require significantly more specialist advice in relation to the problem, or would advise their client not to purchase such an investment.

Those who did value the property (only 25%) made substantially greater upwards adjustments to the yield figures which they adopted. (One respondent added 5% to the reversionary yield in order to reflect this increased risk, whereas the other respondents added 2%.)

Question (3)

Your client asks for an analysis of worth of a freehold retail investment, which lies upon soil which is contaminated with non-migratory heavy metals.

You are given the following information:

* *The party responsible for the original contamination is the current freeholder (the developer) and the tenant is not liable at all.*
* *The current rent passing is £180 000 pa, with 4 years until the next rent review. The current lease is 19 years unexpired (5 year reviews).*
* *An estimate of present day remediation costs (by environmental consultants) to clean the site to a ' suitable for use' level has been given as £330 000.*
* *It is anticipated that remediation will take 1 year to complete, during which time the shop will remain closed and no rent will be received. The lease terms enable the landlord to enter the property to implement a clean-up, and there are no grounds for tenant's compensation.*

* *Legislation demands remediation in 4 year's time (at the next rent review). For the 2 years after remediation has been completed monitoring will have to be undertaken. This will be at a cost of £1 500 pa (in current day terms).*
* *All-risk-yields and full rental values on similar uncontaminated properties stand at 10% and £200 000 pa respectively.*
* *Assume both cost inflation and rental growth at 2% pa average.*

It should be noted that in reality, as was indicated by some of the respondents, it is unlikely that legislation would demand remediation in 4 years' time and that monitoring would be required, were contaminants deemed to be non-migratory. However, this does not detract from the purpose of the valuation scenarios, as they were intended to assist in the development of approaches which could be applied to the valuation of contaminated land and property.

Results

60% of respondents to this question applied a DCF method in order to give a net present value figure, thereby indicating the worth of the investment. The remaining 40% used a conventional open market value (OMV) term and reversion method in order to produce a market valuation.

There was a significant disparity over the choice of equated yield that was applied by those respondents who used a DCF approach: figures ranged from 9% to 12%.

It was ascertained that an 'analysis of worth' would be judged against the open market value (OMV) in the decision to purchase by an investor. The best practice answer therefore used both an open market term and reversion valuation and a DCF appraisal.

For the DCF calculation, an equated yield of 11.5% was used [1]. This was firstly composed from a long-dated gilt yield of 8.5% (which can be seen as a risk free investment in money terms). To this a premium of 2% was added, to account for the additional risks which are experienced in property investments (including those of tenant default and illiquidity). Finally, an additional premium of 1% was added for stigma, giving 11.5%.

Best practice answer

Using this yield of 11.5% gives a positive net present value (NPV) above the OMV of some £59 000. Therefore, the investment is worth at least the OMV (£1 467 000) if an investor wishes to attain an 11.5% return on his investment. However, an equated yield of 12% would give a negative NPV (below the OMV) and as such, the investor would not attain a 12% return on their initial outlay, were they to purchase the investment.

As with the previous two answers, financing costs were also included for remediation, and the costs of environmental investigations and the preparation of an LQS were also incorporated into the calculation.

Qu. 3			
Term Income	£180 000		
YP 4yrs @ 10%		3.169	£570 420
Reversionary Income	£200 000		
YP Perp@ 10.5%		9.52	
PV £1 %yrs @ 10.5%		0.607	£1 115 728
Total			£1 726 148

Less:

Costs of intrusive study (pre-acquisition)		£ 22 500[2]	
Remediation costs	£330 000		
Finance costs (6 months @ 10%)	£ 16 500		
	£364 500		
PV £1 4yrs @ 10.5%		0.671	£232 502
Monitoring costs in year 5(4)	£ 1 500		
PV £1 5yrs @ 10.5%		0.607	£911
Monitoring costs in year 6(4)	£ 1 500		
PV £1 6yrs @ 10.5%		0.549	£824
LQS (Preparation/Report)	£ 5 000[5]		
PV £1 7yrs @ 10.5%		0.497	£2 485
			£259 222
Value			£1 466 926
		Say	**£1 467 000**

CONCLUSIONS

As part of the conclusion, and as a summary of the main issues, a 'flow' diagram, showing the suggested route to the valuation or the calculation of worth of contaminated investment properties, is given in Figure 2.

A first stage in the valuation process is the assessment of the environmental risk which relates to a property. Depending upon the levels and extent of PI insurance and experience in relation to environmental issues, this may either be conducted solely by the valuer, or as part of a 'team approach' basis by liaising with an environmental consultant.

Qu 3(B) DCF calculation

Yrs	Cur.Rent/FRV	Inflation @ 2%	Projected Rent	Costs	Inflation @ 2%	Projected costs	Net cashflow	YP @ 11.5%	PV @ 11.5%	NPV
0				£22 500		£22 500[2]	£22 500			-£22 500
1-4	£180 000						£180 000	3.070		£552 530
4	Remediation costs			£330 000	1.082	£375 063[3]	£375 063		0.647	-£242 663
5	VOID (remediation carried out[3])									
5	Monitoring[4]			£1 500	1.104	£1 656	£1 656		0.580	-£961
6	Monitoring[4]			£1 500	1.126	£1 689	£1 689		0.520	-£879
7	LQS[5]			£5 000	1.148	£5 743	£5 743		0.467	-£2 681
6-9	£200 000	1.082	£216 486				£216 486	3.070	0.580	£385 603
10-14	£200 000	1.195	£239 019				£239 019	3.650	0.375	£327 519
15-19	£200 000	1.32	£263 896				£263 896	3.650	0.218	£209 828
20-Perp	£200 000	1.457	£291 362 [6]				£291 362	8.696	0.126	£320 268
										£1 526 063
									Say	£1 526 000

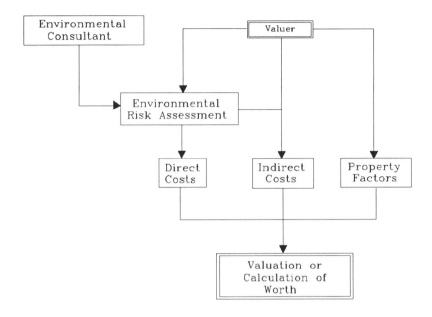

Figure 2 A valuation approach for contaminated land and property

This assessment will provide an estimation of direct costs, including those for remediation and monitoring which may need to be incurred in relation to the property. These costs will form an input into the valuation or calculation of worth. The valuer must then use his knowledge and experience of the market in order to determine the likely indirect or 'stigma' costs and the importance of any 'property factors' which may impact upon the valuation.

These three inputs are then combined within a valuation method or a calculation of worth. It has been shown that the 'cost to correct approach' can offer a suitable means by which these factors can be used to produce an open market valuation. However, care must be taken, both in terms of the comparative information that is used as an input and in the way in which stigma is incorporated into the valuation.

This paper has shown that stigma may be integrated into such a valuation by making an appropriate adjustment to the ARY figure. However, it must be emphasised that such an adjustment is highly subjective, and that the figures which are presented here are only intended to indicate the type of approach which may be adopted.

DCF techniques can similarly be applied to calculate the worth of contaminated properties. Indeed, the flexibility of DCF, particularly in relation to incorporation of time-dependent costings and the explicit nature of such techniques, suggests they should be considered as an integral part of the valuer's outlook. In particular, in relation to the valuation of contaminated properties, they may reveal inconsistencies within traditional methods, for example those concerning implicit assumptions made as to growth. However, any adjustments made for stigma will also be highly subjective, even when using such explicit methods.

Implications for valuers and the profession

Although respondents to some extent covered themselves with protective assumptions in their responses, it was clear that there was much uncertainty and therefore significant disparity amongst the answers which interviewees provided for the valuation scenario questions, and in the levels of environmental knowledge which they demonstrated. Some answers were highly confused, whereas others were well thought out and logical. There is therefore a distinct need to widen the debate and disseminate the knowledge that does exist more evenly throughout the valuation profession.

It is therefore suggested that more education for valuers in this specialist area is vitally needed, so that they have an appropriate knowledge of basic environmental risks and relevant legislation and suitable valuation methodologies.

This may be achieved partly through the improvement of the current professional guidance. For example, interviewees suggested that guidance within the new RICS Appraisal and Valuation Manual ('Red Book') could be improved to incorporate the following points:

A 'Property Observation Checklist', which valuers could use to assess the basic environmental risks of a site or property whilst conducting a 'walkover' survey (Colangelo and Miller (1995)). Such an approach has proved worthwhile in the USA, and has also eased PI insurance problems. (In this respect, PI insurers were much more content for valuers to follow a set procedure in assessing environmental risk.) In the USA this checklist is voluntary and offers a 'limited scope analysis'. Only visual on-site observations are recorded: the intention is to identify possible environmental factors that could be observable by a non-environmental specialist: for example, evidence of past or present on-site industrial activity, stained soil, distressed vegetation, tanks, vent pipes indicating underground storage tanks. The user of such a checklist is, however, reminded that the appraiser is not an environmental specialist and so the checklist is used as a way of assessing whether an environmental specialist is needed.

Advice could be offered on the ways in which different valuation methods may be adapted and the underlying factors which should be borne in mind by a valuer in such a position; and

Stigma A more specific definition of stigma and how it may be dealt with in valuation would also prove useful. A useful working definition from the USA could provide a starting point (Colangelo and Miller, 1995)

'...The negative impact that results from public perception that environmental contamination is permanent and represents a continuing risk even after environmental cleanup has been completed. Also refers to the negative result on property values that occurs in properties in close proximity to contaminated sites...'

The new Red Book will become a mandatory document on 1 January 1996. The contamination guidance within the manual replaces existing guidance which was previously contained within VGN11. The revisions to VGN11 fall under the following sections within the new manual:

Practice Statement 2 - Conditions of engagement
Practice Statement 4 - Definitions of Bases and Assumptions
Practice Statement 6 - Inspections and Material considerations
Practice Statements Appendix 3 - Typical Paragraphs in Valuation Reports (details
        caveats which may be attached to valuation reports)
Guidance Note 2 - Environmental Factors, Contamination and Valuation

These revised guidelines appear to go some way towards addressing those issues which were raised by the interviewees. It will be interesting to assess the profession's reaction to these revised guidelines.

The degree of subjectivity and uncertainty in relation to the valuation of contaminated land and property may also be eased by the establishment of a contaminated land 'database' (as suggested by Syms (1995)). Such a database could also provide comparable information as to remediation costs and stigma costs. However, it must be emphasised that, in practice, such a 'contaminated comparables' database could be very difficult to compile and maintain, due to fears over confidentiality.

Finally, it is clear that more research is needed in the area of contaminated land valuation. For example, research into ways into which indirect 'stigma' costs may be measured and incorporated into valuations or calculations of worth to improve valuation accuracy is particularly needed.

Much remains to be done if valuers are to deal with these issues effectively and provide accurate and meaningful advice to clients. Environmental risk is important but must not be dealt with in isolation. In the words of a lender, interviewed in our research, the risks which contamination present can be viewed as a:

'...landscape of risk that has been full of very very familiar features for many years. Now, all of a sudden, we get "environment" coming along and posing new risks, which become new features in that landscape of risk. You can separate out these 'new features', but it must always be remembered that they also affect all of the other features.'

ACKNOWLEDGEMENTS

The authors would like to thank the following for their invaluable help and advice in the production of this paper:

Elisabeth Culbard	Laurence Johnstone
Aspinwall & Co.	Rogers Chapman, Chartered Surveyors
Ian Morgan and David Law	Philip Wilbourn
Allsop & Co	Philip Wilbourn & Associates

They would also like to acknowledge the generous funding of the research by the Pat Allsop Charitable Trust.

BIBLIOGRAPHY

Barrett, H. (1995). Environment Bill - new controls. *Estates Gazette,* January 28 146-147.

Chalmers, J.A. & Roehr, S.A. (1993). Issues in the valuation of contaminated property. *Appraisal Journal.* January, 28-41.

Colangelo, R.V. & Miller, R.D. (1995). *Environmental Site Assessments and their Impact on Property Value: The Appraiser's Role.* Appraisal Institute, Chicago.

Confederation of British Industry (CBI) (1994). *Tackling Contamination - Guidelines for Business to Deal with Contaminated Land.* CBI, London.

Department of Environment (DoE) (1994a). *Paying for Our Past.* DoE, London.

Department of Environment (DoE) (1994b). *Framework for Contaminated Land.* DoE, London.

Dixon, T.J. (1995). *Lessons from America: Appraisal and Lender Liability Issues in Contaminated Real Estate.* College of Estate Management, Reading.

Dixon, T.J. & Richards, T.O. (1995). Valuation lessons from America. *Estates Gazette.* July 22 110-112.

Eliot-Jones, M. (1995). *Bixby Ranch: Some Observations on Plaintiffs Expert's Appraisal of Post-Clean-Up Stigma.* Foster Associates, San Francisco.

Environmental Assessment Group (EAG) (1993). *Valuing Contaminated Land - A Need for Shared Experience.* EAG/Montagu Evans, London.

Hillier Parker (1994). *The Attitude of Property Investors.* Hillier Parker,London.

Hillier Parker (1993). *The Impact of Environmental Issues on Asset Value.* Hillier Parker, London.

House of Commons Environment Committee (1990). *Contaminated Land: Volumes I, II and III.* HMSO, London.

Jenkins, S. (1995). DoE attacks sale of polluted gas site. *Estates Times.* 3 February.

Lizieri, C., Palmer, S., Charlton, C. and Finlay, L. (1995). *Valuation Methodology and Environmental Legislation:A Research Project for the RICS Education Trust.* City University Business School, London.

MacRae, J. (1994). Funds 'too negative' about polluted land. *Estates Times*. 14 October.

Mundy, B. (1992a). The impact of hazardous and toxic material on property value: revisited. *Appraisal Journal*. October, 463-471.

Mundy, B. (1992b). The impact of hazardous materials on property value. *Appraisal Journal*. April, 155-162

Mundy, B. (1992c). Stigma and value. *Appraisal Journal*. January, 7-13.

National Westminster Bank (1994). *Environment Report*. NatWest Bank, London.

Parliamentary Office of Science and Technology (1993). *Contaminated Land*. HMSO, London.

Patchin, P. (1988). Valuation of contaminated properties. *Appraisal Journal*. January, 7-16.

Patchin, P. (1991). Contaminated properties-stigma revisited. *Appraisal Journal*. April, 167-172.

Patchin, P. (1994). Contaminated properties and the sales comparison approach. *Appraisal Journal*. July, 402-409.

Richards, T. (1995). *A Changing Landscape: The Valuation of Contaminated Land and Property*. The College of Estate Management, Reading.

RICS (1993). Valuation Guidance Note 11 - Environmental Factors, Contamination and Valuation. *The Manual of Valuation Guidance Notes*. RICS, London.

RICS (1995a). *RICS Valuation and Appraisal Manual*. RICS, London.

RICS (1995b). *Land Contamination Guidance for Chartered Surveyors*. RICS, London.

Sheard, E.M. (1993). Valuing contaminated land: asset or liability. *Land Contamination and Reclamation*. 1(1): 9-15.

Sheard, E.M. (1992). Valuation of contaminated land: current theory and practice. *Journal of Property Valuation and Investment*. **11**: 17-27.

Stephenson, K. (1993). Valuation problems. *Solicitors Journal*. 12 February, 130-131.

Syms, P.M. (1995). 'Environmental Impairment: An Approach to Valuation'. Presented at *The Cutting Edge, RICS Research Conference*.

Tromans, S. and Turrall-Clarke, R. (1994). *Contaminated Land*. Sweet and Maxwell, London.

Wilson, A. (1994). The environmental opinion: basis for an impaired value opinion. *Appraisal Journal*. July, 410-423.

Wilson, A., Ramsland, M., Wilhemy, T., and Groves, R. (1993a). Ad valorem taxation and environmental devaluation Part I: An overview of the issues and processes. *Journal of Property Tax Management*. Summer, 1-32.

8 Planning in the Coastal Zone: A Comparison Between England/Wales and Sweden

JANE TAUSSIK
University of Portsmouth, Portsmouth, UK

INTRODUCTION

'In coastal regions and in the sea, all human activities in one way or another affect the possibilities for alternative use and/or further exploitation of the natural resources.' (SIDA,. 1993)

The purpose of this paper is to examine the contribution town and country planning may make to managing the exploitation of coastal resources and whether different legislative and administrative frameworks cause planning to be more or less effective. It suggests that coastal planning needs to operate within a wider context of land and water planning than it has traditionally and that operating in this way will stretch the boundaries of traditional planning knowledge, requiring much more co-operation and integration with other major land and water interests than currently exists.

To maritime nations the coast represents an enormous resource. Historically, the sea offered relatively easy transport, the basis for exploration and trade. Economically, the resources of fisheries, energy and minerals have been exploited directly while the water mass itself provided opportunities for the dilution and dispersion of waste materials. Socially, it provided opportunities for recreation and relaxation, whether the quiet enjoyment of the walker or artist or the active participation of the sailor or fisherman. Such recreation may be enhanced by the richness of coastal flora and fauna, the diversity of which reflects the level of adaptation of natural species to specific environmental conditions. Strategically, the coast represented vulnerability, a location which, in the national interest, must be defended. In many places, for many countries, such defence has been not only the defence of the nation from other nations but also the defence of the land from the sea.

This wide range of uses of the coast (see Figure 1) reflects the elements that comprise the coast: the land, the water mass of the sea, the inter-tidal zone and the sea bed. It is, therefore, more appropriate to consider a 'coastal zone' where coastal land and coastal waters strongly influence each other than to consider the coast as the line

The Environmental Impact of Land and Property Management, Edited by Yvonne Rydin
©1996, The Royal Institution of Chartered Surveyors

where land meets sea.

Consideration of the ways in which the coast is valued suggests that conflict exists between different user/interest groups. For example, a coast protection scheme for a cliff top settlement may deny replenishment material to another stretch of coast, so that beaches are lost to the sea and the tourist industry is lost to the local economy (Lee, 1995). Recently, a public inquiry heard arguments concerning possible pollution and landscape implications of exploiting coastal mineral resources through a super-quarry development (Johnson, 1994; Burton, 1994). Currently, the Ministry of Defence's move of a naval practice firing area to a location off the Cornish coast is causing concern about the biotic resources and recreational interests in the locality (Hornsby, 1995; Purves, 1995). These UK examples are typical of the range of conflicts which occur worldwide.

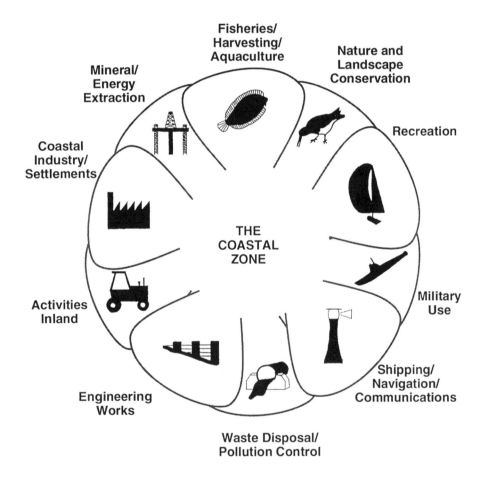

Figure 1 Major uses in the coastal zone

COASTAL PLANNING IN ENGLAND AND WALES

In many countries, resolution of such conflicts is impeded because of the sectoral nature of responsibility for land and marine resources. This is particularly true in England and Wales where different bodies are concerned with land and sea based resources, and where the boundaries of responsibility vary between bodies (see Figure 2). To overcome this problem, it has been proposed that integrated coastal zone management be adopted nationally (Gubbay, 1990; RSPB, 1993), to follow practice in countries like New Zealand (Murch, 1993) and the USA (Carter, 1988). Such an approach requires that current and future needs of all user and interest groups be considered in the context of the resources available and the demands/values placed on them, so that decisions can be made which protect the long term quality of coastal resources. While the House of Commons Environment Select Committee (1992) recommended such an approach, the Government has chosen not to pursue it (DOE, 1992a). For the meantime, therefore, existing spheres of responsibility in the coastal zone in England and Wales remain.

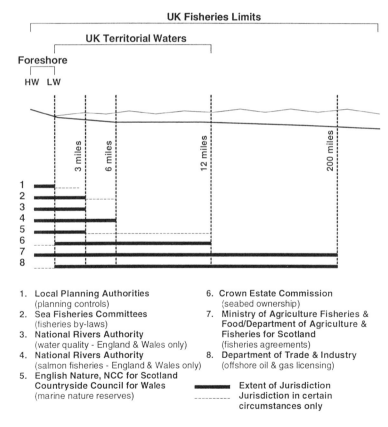

Figure 2 The extent of jurisdiction of some UK coastal organisations (after Gubbay, 1990)

The town and country planning system, with its combination of forward planning and control of development, provides a well established mechanism for resolving conflicting demands in the use and development of land and property resources (Brindley et al., 1989) which can make a significant contribution to coastal zone management. It is operated by local planning authorities (LPAs) who have the duty of undertaking planning within their administrative areas. The seaward boundary of coastal local authorities is, normally, the low water mark. In the context of the coastal zone identified above, LPAs, therefore, have responsibility over only the land and the inter-tidal zone and not over the sea or the sea bed. This limits the contribution planning can make to coastal zone management.

The division of responsibilities for coastal developments between a range of organisations creates a series of anomalies. For example, to construct a marina involves a host of permissions concerned with land, the sea bed and the water mass (Taussik, 1995a). To extract sand and gravel from a land site requires planning permission from the LPA. To extract sand and gravel from the sea bed requires a licence from the Crown Estates, a body which gains financially from its granting (Hollingsworth, 1994). Various criticisms may be levelled at this system: it is complex and time consuming; cooperation between relevant bodies such as the Ministry of Agriculture Fisheries and Food, the Department of Trade and Industry and English Nature, may not happen; the decision making process is not publicly accessible or accountable; the objectivity of licensing bodies may be questioned even though mechanisms exist to obtain independent government views (The Government Review Procedure).

Various organisations and institutions (e.g. RSPB, 1993) have been pressing for the planning system to be extended to the limit of the territorial sea or, perhaps, to a closer limit. This would allow the decision making process to be more transparent, would allow for third party representation to be made (or more easily made) and would make the decision making democratically accountable and removed from those with direct interests in the outcomes. It would also enhance substantially the contribution that planning could make to coastal zone management.

Such a recommendation was made to the government (House of Commons Environment Select Committee, 1992) but, again, was not accepted (DOE, 1992a). Planning continues to be limited to land above the low water mark and its direct contribution to management in the coastal zone thereby remains limited, although the contribution to coastal management of certain LPAs, like Hampshire County's and Sefton Borough's, is exemplary.

However, planning policy guidance for the coast prepared by the Secretary of State for the Environment (DOE, 1992b) recognises the importance of the concept of 'the coastal zone' and the inter-relationship of the elements which comprise it, although the water surface, the water mass and the sea bed are not separately differentiated. It also recognises the limitation of dealing with coastal issues, which may relate to the coastal cells which reflect coastal processes (Carter, 1988) or to complete estuaries bisected by county and/or district boundaries, by administrative areas. Therefore, it requires LPAs to take into account the effect of on-shore development on off-shore areas in making planning decisions and advises coastal LPAs to work closely with other authorities on coastal issues.

In spite of this guidance, which is acknowledged to be relatively new, it is suggested that LPAs do not actively consider all elements of the coastal zone. For example, analysis of development plans suggests that increasing prominence is given to coastal issues and coastal policies. However, very few LPAs consider marine elements in their development plan policies, either for the water mass or the sea bed (Taussik, 1995b). A 'land by the sea' interpretation of 'the coast' can be inferred. It is suggested that LPAs' long tradition of dealing with land use and development, combined with lack of control of marine use and development, has made them vulnerable to this narrow interpretation of 'the coast' which will take considerable effort to broaden.

COASTAL PLANNING IN SWEDEN

Like England and Wales, Sweden is a nation with a long maritime tradition. A higher proportion of the Swedish population live along the coast than that of the population of England and Wales, reflecting the relative ease with which coastal resources could be exploited compared to those inland. While the problems of coastal Sweden differ from those in England and Wales, the same pattern of uses and interests and the same areas of conflict emerge. For example, Sweden has few of the 'bucket and spade' resort problems of Engalnd and Wales but it does have enormous problems with the hundreds of thousands of summer houses scattered on its shores, often at the water's edge (Segrell and Lundqvist, 1993). Both countries have major industrial complexes like oil refineries which require coastal locations, usually in areas of high natural value. In both countries concern is expressed about the environmental impact of super-ferries, of wind generated energy and of aquaculture.

Like England and Wales, as Figure 3 shows, Sweden has a considerable body of legislation, operated by a range of organisations, for coastal land, the sea and the sea bed. Like England and Wales, there is no specific legislation related to coastal zone management although the Swedish Marine Resources Commission (now abolished) established the principles of coastal zone management (1989). It is recognised that physical planning has had an important role for many years as a 'coordinating instrument for issues involving the use of land resources in coastal areas' and it is suggested that current approaches to coastal and marine management had developed from it (Carlberg et al., undated). The current legislative framework is provided by the Planning and Building Act 1987 which states:

'Section 1 Article 2: It is a municipal responsibility to plan the use of land and water areas' (Boverket, 1993a)

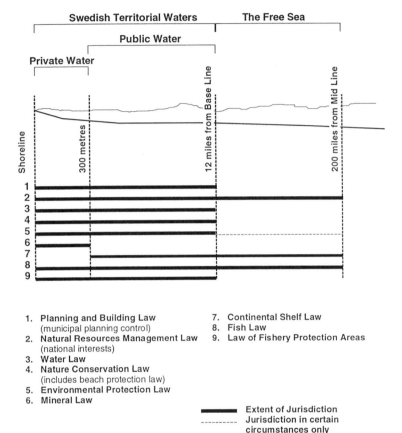

Figure 3   The extent of some Swedish laws in coast and sea areas (after Boverket, 1994)

Such planning must operate within the context of the Natural Resources Management Act 1987. This allows for the identification of areas of national importance because of their scientific or historic value or their value for industry or defence and states:

> 'Section 1 Article 1: Land, water and the physical environment in general shall be used in a manner that encourages good long-term management from ecological, social and economic viewpoints.' and

> 'Section 2 Article 1: Land and water areas shall be used for that or those purposes for which they are most suited with regard to their nature and location as well as actual requirements. Precedence will be given to that use which encourages good management from the public's viewpoint.' (Boverket, 1993b)

This latter section is repeated in the Planning and Building Act, 1987.

These two acts, therefore, provide a coastal planning framework rather different from that in Engalnd and Wales. Unlike in England and Wales where the Town and Country Planning Act 1990 refers to land, in Sweden, the use of both land and water must be planned. By 'water' is meant rivers, lakes and ground waters as well as coastal waters. Together, these contribute to a water system which finishes at the sea, to where all precipitation in Sweden will, eventually, drain (Arnemo and Adolfsson, 1995). The Propositions, which pre-date enactments but form part of legislation, state (SOU, 1985) that municipalities' responsibilities extend to the limit of the territorial sea, i.e. to 12 nautical miles out from the baseline (Grip, 1992). It would appear, therefore, that the Swedish planning system is much better equipped, legally, than the system in England and Wales to deal with conflict in the use and development of its coastal zone.

The municipalities (kommuns) have almost complete autonomy over planning matters subject to their recognition and protection of resources recognised as being of national importance. Such interests are identified by national boards, supervised by county administrations (länsstyrelsen) which are central government bodies, operating at regional level, providing information and ensuring that nationally important resources and the interests of adjacent authorities are protected (Swedeplan, 1991).

Each municipality must produce a comprehensive plan (översiktsplan) for its area to show how land and water areas are utilised and how development should take place (Planning and Building Act: Section 1 Article 3). These plans are an integral part of the planning system and, because they interact with a wide range of sectors dealing with land and water, they have a coordinating role to play in coastal zone management (Ackefors and Grip, 1994). These plans are advisory but detailed development plans (detaljplan) should conform to them and discussion about development proposals refer to them (Swahn et al., 1995). The control of development takes place through the detailed development plan, without which no development of any scale can proceed and which, itself, provides the right for conforming development to be undertaken. Building permits are required (Fredlund, 1991; Johnson and Sterner, 1994).

The requirements for planning to include land and water areas and for comprehensive plans are relatively new, being introduced by the legislation of 1987. At the same time, planning was made more decentralised, with the municipalities' role enhanced at the expense of central government's National Land-Use Planning which had operated for some 20 years (Byland, 1994; Westman, 1991).

THE EXTENSION OF PLANNING OVER WATER

There are several reasons for the extension of planning over water areas. Territorial Sweden includes large areas of water: there are some 100 000 lakes (The Swedish Institute, 1995) ranging from the enormous lakes of Vänern and Vättern and Stockholm's Lake Mälaren to small lakes dotted through the landscapes and townscapes of the country; parts of its coast are archipelagic and include large islands with much settlement and economic activity and tiny, bare, rocky islets. The Stockholm archipelago, for example, includes some 25 000 islands (*Sweden 1994*, 1994). These water areas, historically, have provided opportunities for economic exploitation (Segrell, 1994) and, hence, the basis of settlements.

Swedish municipalities have particularly broad responsibility concerning water: water supply; the treatment and discharge of waste water; control over discharges into water; water based recreation (Månsson, 1992). This, and the substantial dependence on surface waters (Malmqvist, 1993), has resulted in much greater awareness of the inter-relationships between water discharges, water quality and water use than, perhaps, is the case where responsibilities are divided and there is a long delay between discharges and water use. The effect of development on water systems is clearly recognised so that "Coastal areas and drainage basins of rivers must be seen as interlinked economic, cultural/societal and ecological systems" (SIDA, 1993). There is support, therefore, for the basic planning unit in Sweden to be the drainage basin or catchment area extending out into coastal waters (Boverket and Naturvårdsverket, 1989; Arnemo and Adolfsson, 1995).

While 93% (Swedeplan, 1991) of the Swedish population lives in urban areas, the urbanisation of Swedish society is a twentieth century phenomenon (Goldfield, 1982). There are powerful pressures to enjoy undeveloped space, to access the natural environment and to fish and hunt which are expressed through the very high ownership of summer houses, mainly on or close to beaches (Segrell and Lundqvist, 1993), and the withdrawal from the cities to the coast/countryside of high proportions of the population for July. These pressures have found expression in certain laws or rights (Segrell, 1994); discussion of the Holiday Act of 1938 identified open air bathing as one of the best counter-balances to urban life; Allemansrätten (Everyman's Right) provides the right of access over the countryside and water (*Sweden 1994*, 1994); and Riparian Law allows for the protection of shores for public access. The most rapid phase of summer house development was through the 1960s and 1970s (Segrell, 1994) when private owners retained their development rights in areas of 'sparse development' (Rudberg, 1991). Summer houses were built on shores, both continuously along the shore and in very remote, exposed locations, including on tiny islets accessible only by boat. Although such development is now controlled, its use represents a very substantial interest in the coastal zone. Because of its scale, it generates substantial environmental problems and is a common coastal issue.

The mid-twentieth century saw industrial demands made on coastal resources: for land for nuclear, pulp, oil and chemical industries and for water as both a raw material and a medium for waste disposal (Segrell, 1994). Often the preferred industrial locations were areas of high natural value, heavily used for recreation. Such conflicts were instrumental in the development of the National Land-Use Planning which operated from 1973 to 1987 (Forsberg, 1992) to resolve the conflict between two government objectives: for efficient industry and for recreation (Segrell and Lundqvist, 1993). It allowed central Government to identify areas of land and water to be protected because of their natural quality or their economic potential, a function now operating through the Natural Resources Management Act 1987.

Following the Free Commune Experiment in the 1980s (Häggroth et al., 1993), the decentralisation of responsibility from central to local government included planning and resulted in the re-establishment of municipality as the primary planning authority. In response to emerging coastal conflicts (Ackefors and Grip, 1994), another experiment in 1982-83 was directed towards establishing the capability of municipalities to undertake water planning, an experiment that was extended to further municipalities in 1987-89.

It is clear from the discussion in *Proposition 1985/86* (SOU, 1985) that two principles had become established: that municipalities should be the primary planning units; and that planning should include both land and water areas to allow environmental thinking in planning decisions. The inter-relationship between land and its use and water areas and the water table and their uses are cited as explanations of the need to integrate water planning with other physical planning. The comprehensive plan is the vehicle in which the municipality must identify how it proposes to deal with land and water areas. In it, the municipality must identify established national interests, which cover 80% of the Swedish coast (Segrell, 1994), and show how they are to be protected as well as identifying other local, public interests which it considers important in determining the future use of land and water.

COMPREHENSIVE PLANS AND THE PLANNING OF THE COAST

The legal framework for coastal planning of land and water areas, the experience from two sets of trial municipalities in water planning and the national interests in the coastal zone identified in National Land-Use Planning all suggest that the planning of Swedish coastal areas should provide an example for other countries.

Recent research into comprehensive plans in coastal areas undertaken by Boverket (the Swedish equivalent to the Department of the Environment) and academic institutions suggests that experience is very mixed. The earliest research (Andersson, 1991) found that only a limited number of comprehensive plans had been produced by the stipulated date (1990), mainly because of limited resources. Available plans tended to concentrate on land use and development with little content relating to, for example, coastal nature resource problems.

While more recent research (Boverket, 1995; Engen, 1995a and forthcoming) has been based on almost complete plan cover, conclusions about coastal planning remain disappointing. Firstly, land and water are not universally included in comprehensive plans and, even where both exist, they frequently generate separate elements of the plan so the intended integration (SOU, 1985) is lacking. Secondly, coastal water and sea areas are frequently not identified as municipal planning areas and are, therefore, not included in the plan. Thirdly, while some municipalities include their archipelagic land and water, relatively few include the full extent of coastal waters (i.e. to the limit of the territorial sea). Fourthly, where municipalities have included coastal water areas, its treatment is much less satisfactory than the treatment of land areas. It also tends to relate only to the water surface and not to include the water column or the seabed.

Finally, the linkages between land use and water quality and its control are not clearly developed in the plans. Although some municipalities have advanced considerably in terms of coastal water planning, generally there is great uncertainty among municipalities about the extent of their planning responsibility and the methods they should use for planning in coastal and sea areas.

Several explanations are offered for this poor response (Boverket, 1995). The time period for producing the first comprehensive plans was short (Byland, 1994) and the requirement to complete them in an election year inappropriate (Totschnig et al., 1995). Planning of water areas was just one of several new areas of responsibility for municipalities when comprehensive plans were introduced and the lack of tradition in dealing with it (a problem exacerbated by the architectural background and age of Swedish planners) meant it received low priority. The information base for this new area of study varied: for some areas on the west coast it was relatively good but for other areas it was very limited. Material available from central government (e.g. Boverket and Naturvårdsverket, 1989) concentrated on land based issues and did not highlight maritime issues, perhaps reflecting the untimely closure of the Swedish Marine Resources Commission shortly after the introduction of the 1987 legislation. There are also reservations about the municipalities' resources and competency to undertake these new responsibilities (Engen, 1995a and b).

In any municipality, much depends on the level of interest and knowledge of individuals, particularly of politicians. For example, in Lysekil, the high profile of physical planning generally, and water planning in particular, owes much to the well established Social Democratic control of the area and individual enthusiasms for planning (Trädgårdh, 1995). In other areas, where such interest is not apparent, water planning receives much lower priority. Many municipalities appear to be unaware of the extent of their planning responsibilities seaward (Boverket, 1995) while others feel there is no need for planning to extend so far because there are no conflicts of use in coastal waters beyond the archipelago (Boverket, 1995; Swahn et al., 1995). As emerging demands suggest considerable benefit in including this wider area, such views may reflect a lack of knowledge. For example, there are environmental problems with super-ferries and the disposal of dredge material (Totschnig et al., 1995), the development of wind farm platforms (Swahn et al., 1995) and the implications of creating nature protection areas.

Boverket's early findings about the treatment of the coast in comprehensive plans (Johnson et al., 1991) prompted it to provide additional guidance to municipalities (Boverket, 1994). *Kust och Hav* includes a contextual overview for coastal land and water planning and guidance to coastal municipalities about the types of information required, supported by examples of good practice from existing plans. Currently Boverket is working to provide additional assistance for water planning. Coastal planners at Boverket (Rönning, 1995) consider that a much higher proportion of the second generation plans will incorporate coastal land and water planning, though there is less certainty about the seaward extent of the planning area. These views seem to have wide acceptance (Swahn et al., 1995; Totschnig et al., 1995).

While the content of comprehensive plans indicates the response to the legislative changes, it does not demonstrate the effectiveness of the plans to assist in the resolution of conflict of use in the coastal zone. The Lysekil comprehensive plan is widely quoted as an exemplar of coastal planning. Lysekil lies to the north on the west coast of Sweden, in an area of considerable landscape and scientific interest including Sweden's only threshold fjord at Gullmarn. With a traditional dependence on fishing and quarrying, more recently its deep water access was responsible for its identification as nationally important for industry. Lysekil was one of the second trial municipalities for water planning when it had the benefit that considerable coastal water information was available because of the location in Gullmarnfjord of Sweden's first marine research station established 150 years ago. Lysekil was also given generous support by its county administration of Göteborgs och Bohuslän and by government bodies. It was, therefore, in a somewhat advantageous position compared to some other authorities in preparing its comprehensive plan.

While the plan extends only to base line and there are some reservations about the information base (Trädgårdh, 1995) and the level of integration between land and water elements (Engen, 1995b), the plan is highly valued in assisting with land and water use decisions. This reflects the politicians' attitudes to planning and the plan, views which are no doubt influenced by the mix of very large scale, potentially damaging industrial development (the Scanraff oil refinery is the largest in Europe) and the high environmental quality of both land and marine environments and the resulting high tourist/recreation use of the area. While other bodies like the Water Courts (Carlberg, undated) use the comprehensive plans and the government may base decisions in matters of dispute on it (Swahn et al., 1995), the view that comprehensive plans are effective in establishing the balance between different interests and in resolving development issues is not universally held (Byland, 1994). Some politicians consider comprehensive plans undermine their power and, while they conform to the requirement to prepare them, do not refer to them in the development process (Arnemo and Adolfsson, 1995). There is no requirement, as there is in Britain, that decisions should normally be made in accordance with the plan (Town and Country Planning Act 1990, Section 54A) and the freedom of municipalities in decision making is a jealously protected principle (Trädgårdh, 1995; Swahn et al., 1995).

Where plans comprise single documents their effectiveness is likely to be enhanced. Many plans comprise a series of documents, often with land and water dealt with separately, without cross-referencing and often appearing ambiguous or contradictory (Swahn et al., 1995). Clearly, the effectiveness of such plans is greatly reduced.

THE NEED FOR INTEGRATED APPROACHES

The point is made in many sources that holistic approaches are required to manage coastal resources (Boverket, 1995; Engen, 1995a; Swahn et al., 1995). Such approaches require the co-operation and integration of different disciplines, of different types of organisation and of different interests (Taussik and Gubbay, 1995; Ackefors and Grip, 1994). The point was also made earlier that catchment areas incorporating coastal areas should be the basic planning units. To plan or manage a catchment area, which may be very large geographically, requires similar co-operation and integration, particularly as the catchment area appropriate from a scientific view is unlikely to conform to the administrative areas for which regulatory, management and providing functions are undertaken.

The wide spread of responsibility of Swedish municipalities for water use and quality were outlined above. To properly plan both land and water use requires integration between planning and environmental protection (Boverket and Naturvårdsverket, 1989; Arnemo and Adolfsson, 1995) and would appear easy to achieve. In some municipalities such integration exists to the benefit of the comprehensive plan. More usually, such co-operation appears lacking and relations between these functions are limited (Arnemo and Adolfsson, 1995; Swahn et al., 1995).

Water moves. It moves downstream and it moves along the coast. Water use problems generated by one municipality become the problem of other municipalities. For water planning, therefore, even more obviously than for land use planning, there is a need for co-operation and integration between neighbouring authorities. There appears to be little inter-municipality planning co-operation though it appears commoner with environmental protection (Totschnig et al., 1995; Engen, 1995b). This may reflect different interpretations of coastal resources with environmental protection officers recognising the coast as a common good or responsibility requiring joint actions and planners viewing the coast as a local fixed resource (Engen, 1995b). It may be that the traditional architectural background of many Swedish planners inhibits their appreciation of the nature of water resources though the newer planning schools are developing more holistic, environmental understanding (Swahn et al., 1995; Engen, 1995b).

The problem of limited knowledge in new areas of responsibility was offered as a partial explanation of the limited coastal water planning in comprehensive plans. In Sweden, this is exacerbated by the self-dependent, autonomous nature of municipalities with their tendency to operate in isolation from each other, though with the close involvement of the county adminstration (Byland, 1994). For example, although Lysekil's coastal water planning work is used to demonstrate Swedish coastal zone management to international visitors (Johansson, 1995; Trädgårdh, 1995), only one of the west coast municipalities outside of Lysekil was aware of it (Engen, 1995a and forthcoming).

Suggestions to reduce these shortcomings include increasing central government's role in planning and in coastal planning particularly (Totschnig et al, 1995), an approach at odds with the Swedish government's policy of decentralisation and the principle of municipal freedom (Swahn et al., 1995; Engen, 1995b). An alternative approach which assists in the dissemination of existing information and the development of cooperative problem solving is to establish groups or networks (Taussik and Gubbay, 1995). In the UK, a large number of coastal groups have been established, of a general nature or related to specific geographical areas or particular coastal problems (NCEAG, 1994), including the Coastal Forum established by the government. Additionally, recognising the international nature of coastal problems, several European networks have been established, for example EUROCOAST (the European Coastal Association for Science and Technology) and EUCC (the European Union for Coastal Conservation). A common purpose of these groups and networks is to develop a more holistic understanding of the coast, its processes and conflicting demands. Such networks related to coastal zone planning or management do not exist in Sweden, either at national or local levels. Although Boverket identifies the need for co-operation to resolve common problems (Boverket, 1994), it has not developed regional conferences on coastal issues although it states that these have been useful for other subject areas (Boverket, 1995). In the context of uncertainty about how to plan and the lack of information, the establishment of local and national groups would increase the confidence of officers and politicians to proceed with coastal planning.

CONCLUSION

As in so many other countries, England and Wales and Sweden are paying increased attention to the resources of their coastal zones and improving the institutional systems which operate to resolve the increasingly obvious conflicts which exist in use and in the expectations of environmental quality. Both have well established planning systems, operating within stable legislative, administrative and democratic frameworks, which can make major contributions to the development of coastal zone management. The extension of planning to consider coastal matters reflects broader changes: originally concerned with the physical requirements of urban areas, both systems have (at rather different times) incorporated rural areas before, more recently, developing approaches to incorporate a broader understanding of environmental issues. In both countries this is relatively new and it is apparent from the example of coastal planning that this is in its infancy and has scope for considerable improvement.

The scale of environmental concern has moved on from concentrating on the externalities generated by specific types of use to the appreciation of the cumulative effects of many small scale changes, suggesting that environmental considerations must underlie all planning decisions. The nature of environmental systems suggests that planning of the coast should be seen as part of a wider water environment - the catchment area extended to include coastal waters - where understanding the inter-relationship between land use and water systems can be used to ensure that use of either does not have negative effects on the other. In this, it is important to deal with land and water together. Compared to the planning system in England and Wales, the Swedish system demonstrates greater appreciation of these inter-relationships although there remains a tendency to treat them as separate elements of the environment.

The coast comprises the land, the inter-tidal zone and the sea. In terms of regulating use and development of these elements, Sweden has created a more integrated framework by extending the planning system to include water use to the limit of the territorial sea. By so doing, it has explicitly recognised the inter-relationships between these elements and their uses. In England and Wales, even though the conflicts of coastal use appear more apparent, the opportunity to do this has not been taken and the complexities of sectoral controls remain, undermining the development of more integrated approaches and environmentally determined decision making. While there is recognition of the inter-relationships in England and Wales, the approach to achieve these objectives is more implicit, developed through policy mechanisms, rather than through the legislative system.

None the less, in England and Wales there have been considerable advances in the generation of strategies to resolve demands for resource exploitation within environmental frameworks. For example, catchment area management plans and coastal management plans for estuaries and harbours are being prepared. These are non-statutory and involve the synthesis of the needs of a broad range of interests identified through inter-sectoral, inter-authority, inter-level groups. While it is too soon to evaluate their success, the voluntary nature of this approach would appeal to Swedish ways of working.

It is clear, in England, Wales and Sweden, that extending the scope of planning into environmental areas generally, and coastal areas particularly, requires the development of more holistic approaches and stretches existing knowledge and information. In the context that such problems cannot be dealt with by small authorities in isolation and that many problems are not unique, there is a need for co-operation and integration between different authorities, different levels of government and different types of interest. There is a need to listen to, and understand, alternative, even opposing, views. The development of local and national networks can assist in the dissemination of knowledge and in problem solving. While neither the membership of coastal groups or the dissemination of information within them in England and Wales would be described as optimal, the number of groups that currently exist suggests that substantial benefits are recognised from the sharing of information, problems and viewpoints. They provide examples which Sweden could follow.

While networks are important, the education system has a vital role in developing understanding of environmental systems. This should happen, not only in the planning schools and environmental studies courses, but also in all courses concerned with land and water, their use and development. Appropriate learning frameworks should be available to current, as well as future, professionals.

Moving into new areas of responsibility takes time. Swedish experience with the 1987 legislation has been disappointing to those anxious to see rapid integration between land and coastal water planning though, on the principle that the 'carrot' may be more effective than the 'stick', the Swedish government's preference for voluntary measures may result in greater effectiveness in the long run. Changes of legislation seldom offer overnight panaceas for problems and the range of reasons for the lack of coastal water planning in comprehensive plans suggest that, were England and Wales to adopt similar legislation, it would take some time before competence in water aspects of coastal planning matched that for land. This provides further support for the development of groups and networks by which coastal zone planning may be promulgated.

Planning of the coast requires broadening of environmental understanding, not only by professionals but also by politicians and the public. Public demands for environmental quality and general environmental awareness can influence community objectives. The Earth Summit in Rio de Janeiro in June 1992 focused on the environment and development and agreement was reached on a number of programmes, including Agenda 21. This is to provide the basis for environmental programmes throughout society. It may provide an opportunity for the development of environmental awareness and education which may influence environmental decision making so that it becomes, to quote current Swedish policy (Engen, 1995b):

'Think global: act local.'

REFERENCES

Ackefors, H. and Grip, K. (1994). *Swedish Coastal Zone Management - A System of Integration of Various Activities*. C.M. 1994/F10, Ref.E. Mariculture Committee. International Council for the Exploration of the Sea. Stockholm.

Andersson, C. (1991). *Översiktsplanering i Kustkommuner. En analys av ÖP 90*. Linköping University, Linköping.

Arnemo, R. and S. Adolfsson. 1995. Interview at Department of Natural Sciences, Kalmar University. 28 July.

Boverket (1993a). *Planning and Building Act*. (English translation). Boverket, Karlskrona.

Boverket (1993b). *Natural Resources Management Act*. (English translation). Boverket, Karlskrona.

Boverket, (1994). *Kust och hav i översiktsplaneringen*. Boverket, Karlskrona.

Boverket. (1995). *Erfarenheter av översiktsplanearbetet. Öp-analys kust och hav - en utvärdering av kustkommunernas översiktsplaner*. Boverket, Karlskrona.

Boverket and Naturvårdsverket, (1989). *Vattnet i kommunal planering*. Naturvårdsverkets Förlag, Solna.

Brindley, T., Rydin, Y. and Stoker, G. (1989). *Remaking Planning: The Politics of Urban Change in the Thatcher Years*. Unwin Hyman, London.

Burton, I. (1994). Harris hearing tests coastal super-quarry concept. *Planning*. 14 October.

Byland, E. (1994). Physical planning according to the planning and building law at municipality level in Sweden: an attempt to turn from top-down to bottom-up planning? in U. Wiberg. *Marginal Areas in Developed Countries*. CERUM Report 1994. University of Umeå, Umeå.

Carlberg, E.C., Percival, M. and Grip, K.. (undated). *Coastal and Marine Management in Sweden*. Swedish Marine Resources Commission, Gothenburg.

Carter, R.W.G. (1988). *Coastal Environments*. Academic Press, London.

Department of the Environment (1992a). *Coastal Zone Protection and Planning. The Government's response to the second report from the House of Common's Select Committee on the Environment.* HMSO, London.

Department of the Environment (1992b). *Planning Policy Guidance. Coastal Planning.* PPG20. HMSO, London.

Engen, T. (1995a). *På den starkares villkor. Kommunal organisering och interkommunalt samarbete om Västerhavets marina miljö.* Slut rapport, Statens Naturvårdsverk Dnr 802-546-91 Fm. Statsvetenskapliga Institutionen, Göteborgs Universitet, Göteborgs.

Engen, T. (1995b). Interview at the Department of Political Science, Göteborgs University. 7 August.

Engen, T. (forthcoming). *Att gå över gränsen (To cross the boundary).* Avhandingsmanus. Statsvetenskapliga Institutionen, Göteborgs Universitet, Göteborgs.

Forsberg, H. (1992). *En politisk nödvändighet. En studie av den fysiska riksplaneringens introduktion och tillämpning. (A political necessity - a study of the genesis and implementation of National Land-use Planning.)* Linkoping Studies in Arts and Science, No. 74. University of Linkoping, Linkoping.

Fredlund, A. (1991). The Planning and Building Act. In *Swedish planning in times of transition.* Swedish Society for Town and Country Planning, Gävle. pp124-125.

Goldfield, D.R. (1982). National urban policy in Sweden. *American Planning Association Journal.* **48**: 1.

Grip, K. (1992). Coastal and marine management in Sweden. *Ocean and Coastal Management.* 18: pp241-248.

Gubbay, S. (1990). *A Future for the Coast? Proposals for a U.K. Coastal Zone Management Plan.* Marine Conservation Society, Ross-on-Wye.

Häggroth, S., Kronvall K., Riberdahl, C. and Rudebeck, K. (1993). *Swedish local Government: Traditions and Reform.* The Swedish Institute, Stockholm.

Hollingsworth, C. (1993). Marine aggregates today and into the 21st Century. *Quarry Management.* August: 17-27.

Hornsby, M. (1995). Navy sails into protest range. *The Times.* 23 August.

House of Commons Environment Select Committee (1992). *Coastal Zone Protection and Planning.* HMSO, London.

Johansson, L. (1995). *Coastal Area Management in Sweden.* SWEDMAR, Gothenburg.

Johnson, A. (1994). What's super about big quarries. *ECOS.* 15 3/4: 35-42.

Johnson, B., Johansson, L. and Rönning, Y. (1991). Öp-Analys Kust och hav - en utvärdering av kustkommunernas översiktsplaner. Arbetsmaterial underlag för diskussion på seminariet om *Kust och Hav.* 18-19 September. Boverket, Karlskrona.

Johnson, B. and Sterner, H. (1994). Interview at Ländsstyrelsen Göteborgs och Bohuslän. 30 August.

Lee, E.M. (1995). Coastal cliff recession in Great Britain: the significance for sustainable coastal management. In M.G. Healy and J.P. Doody, *Directions in European Coastal Management*. Samara Publishing Ltd., Cardigan.

Malmqvist, Y. (1993). Natural Links: A sustainable network hydropolis the role of water in urban planning. *Proceedings of the international UNESCO - IHP Workshop. Wageningen 1993 - Emscher region*. Boverket, Karlskrona.

Månsson, T. (1992). *Ecocycles. The Basis of Sustainable Urban Development*. Environmental Advisory Council, Ministry of the Environment and Natural Resources, Stockholm.

Murch, H. (1993). Antipodean learning curve on coastal planning tide. *Planning*. 3 September.

National Coasts and Estuaries Advisory Group (NCEAG). (1994). *Directory of Coastal Planning and Management Initiatives in England*. National Coasts and Estuaries Advisory Group, Carmarthan.

Purves, L. (1995). This green and pleasant naval practice firing area. *The Times*. 23 August.

Rönning, Y. (1995). Interview at Boverket. 25 July.

Royal Society for the Protection of Birds (1993). *A Shore Future*. Royal Society for the Protection of Birds, Sandy, Bedfordshire.

Rudberg, E. (1991). From model town plans to municipal planning guidelines in Fredlund, A. *Swedish planning in times of transition*. Swedish Society for Town and Country Planning, Gävle. pp101-123.

Segrell, B. (1994). Accessing the attractive coast - conflicts and co-operation in the Swedish coastal landscape during the 20th century. Paper presented at the IBG Rural Geography Study Group *Conference on Accessing the Countryside*. Nottingham. 21-22 September.

Segrell, B and Lundqvist, J. (1993). The attractive coast - context for development of coastal management in Sweden 1930-90. *Scandinavian Housing and Planning Research*. **10**: 159-176.

Statens Offentliga Utredningar (1985). *Proposition 1985/86:1 Ny plan-och bygglag* (Government Bill for the new Planning and Building Act). Statens Offentliga Utredningar, Stockhom.

Swahn, B., von Platen, B. and Erlandsson, M. (1995). Interview at Blekinge Länsstyrelsen. 26 July.

Sweden 1994. (1994). Swedish Travel and Tourist Council, London.

Swedeplan (1991). *Planning in Sweden*. The National Board of Housing, Building and Planning, Karlskrona.

The Swedish Institute (1995). *Fact Sheets on Sweden*. The Swedish Institute, Stockholm

Swedish International Development Authority (SIDA)/Natural Resources Management Division (1993). *Marine and Coastal Resources*. Swedish International Development Authority, Stockholm.

The Swedish Marine Resources Commission (1989). *The 1989 Proposed Overall Programme for Swedish Marine Resources Activities in the 1990's. A Summary.* The Swedish Marine Resources Commission, Stockholm.

Taussik, J. (1995a). Planning and the Provision of Marine Recreation Facilities. In T.J. Goodhead and D.E. Johnson. *Coastal Recreation Management.* E. and F.N. Spon, London.

Taussik, J. (1995b). *Development Plans in Coastal Areas.* Working Papers in Coastal Zone Management, No. 13. Centre for Coastal Zone Management University of Portsmouth, Portsmouth.

Taussik, J. and Gubbay, S. (1995). *Networking in Integrated Coastal Zone Management.* Paper presented at Coastal Zone '95, Conference of EUROCOAST-Ukraine, Odessa. September.

Totschnig, Anéer, A.G. and Wallin, M. (1995). Interview at Länsstyrelsen i Stockholms län. 1 August.

Trädgårdh, C. (1995). Interview at Lysekils Municipality. 3 August.

Westman, B. (1991). Devolution of power to local authorities. In A. Fredlund. *Swedish planning in times of transition.* Swedish Society for Town and Country Planning, Gävle. pp145-151.

9 The Greening of Property Education

A.R. GHANBARI PARSA
South Bank University, London, UK

INTRODUCTION

Environmental issues can no longer be considered detached from the domain of chartered surveyors. A number of recent initiatives by the RICS have helped raise the general awareness of its members on environmental related issues. The launch of 'Land Contamination Guidance for Chartered Surveyors' (RICS, 1995a) and 'Environmental Management and the Chartered Surveyor' (RICS, 1995b) have helped to remind members of environmental issues and their impact on the property profession. Roy Swanston the RICS President (1994-1995) has reminded members that

> *'The property world is far from immune from the impact of increased environmental awareness. If chartered surveyors are to retain their pre-eminent position in the management of land and property they must tackle environmental issues head-on'* *(RICS, 1995b).*

This is further pursued by the current RICS President, Simon Pott (1995-1996), who underlined the importance of the environmental issues and the reaction of the profession to this challenge in his inaugural address to members. By recognising this as a potential area to be exploited by the profession he argued that,

> *'The asset is that we are quintessentially the profession of the physical environment as it is experienced by the vast majority of our fellow human beings. All our efforts are directed to making land, buildings and physical infrastructure work better for people. Yet have we as individuals really made this asset work in our favour? And has the RICS centrally been determined enough in promoting our basic green credentials? A resounding 'No' is the answer to both questions'* *(Pott, 1995).*

Yet chartered surveyors are involved in many aspects of the land development process. Chartered surveyors play an important role in the property development process in their capacity as investment and development advisers, having a direct impact on design specification and client requirement. Furthermore, as valuers they will need

The Environmental Impact of Land and Property Management, Edited by Yvonne Rydin
©1996, The Royal Institution of Chartered Surveyors

to be capable of knowing how to value contaminated land and property assets. Recent research by Davies (1993) highlights the lack of a viable methodology for valuation of contaminated land. The insurance community and creditors are also applying rigid restrictions on potential liabilities arising out of environmental issues (Parsa, 1993). The role of surveyors further extends beyond the development and valuation of property to property management. In their capacity as property managers, they need to be aware of potential tenant risk in different types of property under their control (Parsa, 1995). Turner (1994) in his research into environmental management systems and property investment confirms the potential liability of tenant risk for property investors, insurers and property managers. Such findings underline the seriousness of the challenge facing the surveying profession as a result of the impact of environmental issues on the property development industry.

Publication of the Toyne Report in 1993 also emphasised the importance of environmental considerations in higher education institutions in the UK. Despite the rapid expansion of academic and industry led research into the interaction between environmental issues and the property development industry, there has been little input into property related courses across the UK. However, there have been a number of courses specifically designed to address issues such as environmental impact assessment (EIA), environmental auditing and management and environmental theory, policy and practice. With the increasing sophistication of environmental issues and their impact on the property industry, the property profession faces serious challenges in order to be capable of dealing with such developments both professionally and intellectually. Faced with such pressure, it is important that the next generation of surveyors are educated and trained in such a way that they will be well prepared for the national and international challenge ahead.

Educators and professional organisations such as the RICS play an important role in shaping the educational requirements of the surveying profession through course validation and professional development programmes. Pressure from the European Union (EU) is not confined to industry regulation. The recent directive under the SAVE Programme requires universities to include environmentally related subjects in their construction and civil engineering courses (Smith, 1995). It is, therefore, pertinent to determine how different interest groups involved in the education and training of chartered surveyors could interact in order to ensure environmental issues are addressed in the overall curricula of property education. There is debate between the educators and professionals concerning which subjects should be included in the education of chartered surveyors, but it is difficult to strike the right balance between the intellectual and professional needs of surveyors.

This chapter will seek to evaluate the attitudes of educators, as well as those of professionals who are involved in the training and education of chartered surveyors. It will provide a possible framework for the inclusion of green issues in property education. The research is based on consultation and detailed semi-structured interviews with heads of major surveying schools, course directors, investors and surveying firms as well as examination of selected course material. It also reviews the recent RICS Guidelines and Recommendations for Course Accreditation and utilises the findings of the survey of skills required by chartered surveyors in the Moohan Report (1993). This research does not attempt to provide a complete picture of either surveying education or the property industry but it draws extensively from recent research on related issues.

THE PRESSURE FOR CHANGE

The challenge fundamentally stems from the increasing amount of national legislation and European directives which address environmental issues. Principal amongst these have been the Environmental Protection Act 1990 and the Environment Act 1995. The plethora of over 360 directives from the EU is also having a direct impact on the property industry with subsequent implications for property professionals. For example, the introduction of the European Directive, Natura 2000, will affect property development near protected habitats for flora and fauna (Hall, 1995). The most important challenge facing the property profession is possibly that relating to the issue of contaminated land and property (Wilbourn, 1995). This places an increased pressure on surveyors as valuers. As noted earlier, this has been recognised by the RICS in its recent guidelines, on 'Land Contamination Guidance for Chartered Surveyors' (RICS, 1995a) and 'Environmental Management and the Chartered Surveyor' (RICS, 1995b). The 1995 Guidelines on Contamination call on surveyors to take a lead in tackling the market's lack of understanding and uncertainty over polluted land but stress that members should limit advice given to areas of their own competence and professional indemnity (*Estates Times*, 1995). The RICS has also been active in encouraging research since the early 1990s by funding a number of important projects on aspects of the impact of environmental issues on the property industry. Lizieri et al. (1995) have examined the extent of the use of different valuation methods and firms' attitudes towards environmental problems. The RICS (1994/95) Environmental Research Programme reported in this volume emphasises the diversity of issues that surveyors face in their professional activity.

Combined with the findings of previous research, these new findings could offer some guidance on how surveyors could best be prepared for the increased environmental challenge in their future professional role. This could be achieved provided there is a systematic feedback of academic and industry led research into the education and training of chartered surveyors. However, much depends on how the different interest groups co-operate. A call for a co-ordinated move on course design and environmental action has come from the Toyne Report which was published in 1993, and entitled 'Environmental Responsibility: an agenda for further and higher education'. The main recommendation amongst the 21 broad recommendations was:

'Examining, validating, accrediting and professional bodies should facilitate - and, where appropriate, require - the inclusion of relevant environmental issues in courses for which they control or influence the curricula' (Toyne, 1993).

The Department of Environment's consultation paper, 'UK Strategy for Sustainable Development in 1993', also emphasises the need for environmental training as part of the wider vocational qualifications programme, with a number of lead bodies involved. The Toyne Report stresses in addition the role of vocational training, but as highlighted by Betts (1994), 'the Confederation of the British Industries (CBI) and Further Education Funding Council (FEFC) Inspectorate, in carrying out their separate and thorough reviews of National Vocational Qualifications, made no mention of any developments in environmental education or of the Toyne Report'. Shirley Ali Khan (1993) also argues that 'environmental education is to be treated as a "purpose" comparable to other action agendas, such as "the production of more scientists" and the "increased use of IT".' Nevertheless a result of the recommendations of the Toyne Report many universities have made great progress in the pursuit of higher standards in environmental education. These range from adopting a comprehensive environmental policy and action statement to, for example, the establishment of an Environmental Responsibility Centre at the University of Hertfordshire. The Council for Environmental Education (CEE) provides a national focus in England, Wales and Northern Ireland, encouraging and promoting an environmental approach to education through the creation of partnerships. CEE works in all sectors of education to influence, develop and disseminate good practice (NATFHE, 1994). The Greening of Higher Education Council is also involved with the development of curricula greening and interaction between environmental research and teaching. Environmental issues also emerged as an explicit theme in commissions 7 and 8 of the FIG Melbourne 1994 Congress.

It could be concluded from the above discussion that there is pressure for change (i.e. the inclusion of environmental issues in the education and training of students irrespective of the type of course which they study) both within and outside the educational and professional bodies. In terms of the environmental education of chartered surveyors, a number of issues need to be considered. These are further explored in the following parts of this chapter.

EDUCATION NEEDS OF CHARTERED SURVEYORS

There are conflicting views on current approaches to and the quality of education and training for surveyors. Fraser et al (1994) discuss the need to achieve the balance of what are termed as 'market needs' in terms of professional practice, personal and managerial skills on the one hand, and the "intellectual requirements" including economics, law, financial mathematics/investment/valuation, planning and building on the other hand. In a separate article, Fraser et al. (1995) express anxiety about the quality of surveying courses and entrants and call upon the RICS to support them in maintaining the standards by insisting on divisionally based accredited programmes. Hence the importance of the role of the RICS in regulating the quality of surveying courses.

The Lay Report (1991) defined a surveyor as follows:

> *'A surveyor is a professional person with the academic qualifications and technical expertise to practice the science of measurement; to assemble and assess land and geographic related information; to use that information for the purpose of planning and implementing the efficient administration of the land, the sea and structures thereon; and to instigate the advancement and development of such practices'.*

Within this broad definition a surveyor could be involved with a number of activities from the determination of the size and shape of the earth, the positioning of objects in space and the positioning and monitoring of engineering works, the determination of boundaries of public and private land, the study of the natural and social environment, the planning and development of property, the assessment of value and the management of property, the measurement and estimating costs of construction works to the production of plans, charts and reports in association with other professionals.

There are currently 53 universities and colleges of further education in the UK and Republic of Ireland providing RICS surveying courses in the UK. Some of these are accredited centres with a portfolio of courses. They offer B-Tech, degree and postgraduate courses on different facets of surveying as outlined in Table 1.

Table 1 Type and number of surveying courses accredited by the RICS

Name of course	Degree courses	Postgraduate courses
Quantity surveying	32	2
Building surveying	23	1
Commercial & residential	32	12
Planning & development	8	2
Rural property surveying	8	4
Minerals surveying	4	-
Hydrographic surveying	2	2
Land surveying	8	12
Marine resources management	4	1

Source: Degrees and diplomas accredited by the RICS October 1994. Numbers relate to courses in relevant subjects. Some courses offer exemption for more than one subject therefore above numbers are not exact.

The RICS has published a number of guidelines for course accreditation and skills development. The 1989 report, 'Future Education and Training Policies', foresaw a shortage in the number of surveying graduates to fulfil employment demand and its main recommendations were the need for:

- Maintaining the pre-eminent position of chartered surveyors as the leading providers of services and advice in relation to land, construction and property,
- Responding to the increasing tendency towards specialisation and the increasing diversity of the surveying activity,
- The creation of a more flexible education structure,
- More positive support for academic institutions,
- The development of routes into membership for holders of overseas qualifications,
- The development of a new career promotion package for introduction in 1990/91,
- The establishment of a postgraduate conversion course by the end of 1990/91,
- A review of divisional tests of professional competence (TPC) by the end of 1992/93,
- The provision of a framework of opportunity for technical support staff, and
- Widening the range of expertise entering the profession through non-cognate graduates with a broader range of acceptable skills.

The above recommendations were made during a period of sustained activity in the property market with the assumption of continued growth over the next few years. The 1991 'Procedures and Guidelines - Course Accreditation' put forward a number of recommendations covering:

- Links between the college and the external world of employers and the Institution at local and national levels,
- The rigour of the course in terms of breadth, integration and depth,
- The scope and locus of the course content in the extent of the philosophy, aims and objectives and employer requirements,
- The identification and achievement of the competencies required in the appropriate professional surveying environment,
- The level of knowledge taught and the depth of understanding expected of students through a student centred learning process,
- Maximisation of the intellectual potential of students through a discrete and normally separate learning process,
- Assessment philosophy and mechanisms compatible with the philosophy, aims and objectives,
- A capacity for self evaluation and critical appraisal which includes a considered response to indicators such as student experiences, employer opinion, external examiner comments and validation of accreditation and HMI reports.

This report was designed to integrate the objectives of the RICS, the education offered in colleges and universities and the requirements of the members and employing organisations which the Institution served. It was hoped that the approach to course analysis, evaluation and accreditation would benefit from the application of a well considered and developed methodology applied on a consistent basis (cited in Moohan, 1993).

The Moohan Report (1993) known as 'Procedures and Guidelines for Course Accreditation Supplementary Advisory Notes: Postgraduate Programmes' provided:

1. a model of performance review to course teams to be used in the process of critical appraisal
2. the requirements of professional practice prior to the Assessment of Professional Competence
3. analysis of areas and ranges of expertise required of the surveyor prior to the Assessment of Professional Competence based on an extensive survey of members of the Institution in professional practice and higher education

The report found that there was:

1. no significant difference of opinion between qualified surveyors in professional practice and higher education,
2. general evidence of a balance of agreement but with significant differences between surveyors by division (except land surveying),
3. a strong balance of disagreement between land surveying and all other divisions,
4. a suggested core of skills or areas of expertise classified as high order, medium order or low order across the surveying disciplines,

The Report highlighted the major issues which need specific attention, and in particular the skills or areas of expertise which need to be developed in the surveyor entering professional practice prior to the Assessment of Professional Competence. Moohan cites major problems with the perception and identification of competencies which are subject to change according to different market conditions, internal and external professional influences, the demand for information technology, business, and financial linguistic skills and what he refers to as the 'current fashion for green issues, manifested in notions of sustainable development'. Undoubtedly employers of new graduates have expectations of the type of education that they receive. The 1992 Report to General Practice Divisional Council underlines the fact that 'many employers of graduates wish them to have developed their knowledge, skills and understanding of an area of work in which they are initially to be employed to a greater extent than is possible if a GP related degree course is followed which has no optional element'. The Moohan Report (1993) provides a useful definition and classification of the skills perceived by members of the Institution to be of high order, medium order or low order as outlined in Table 2. Of particular interest are the findings of the survey in terms of the hierarchy of common skills for different divisions. Table 3 outlines the areas of expertise identified by the General Practice Surveying division. It is interesting to note

Table 2 The classification of skills required for surveying courses

	Type of skill	Definition
High order	Non surveying specific	Skills or areas of expertise highly desirable or essential to the surveyor entering professional practice prior to the Assessment of Professional Competence
High order	Surveying specific	Skills or range of areas of expertise requiring an understanding of principles and professional practice with an ability to critically appraise, evaluate and develop principles and practice prior to the Assessment of Professional Competence
Medium order	Non surveying specific	Skills or areas of expertise useful or desirable to the surveyor entering professional practice prior to the Assessment of Professional Competence
Medium order	Surveying specific	Skills of areas of expertise requiring an understanding of principles and practice prior to the Assessment of Professional Competence
Low order	Non surveying specific	Skills or areas of expertise unnecessary or of minimal value prior to the Assessment of Professional Competence
Low order	Surveying specific	Skills or areas of expertise requiring a general awareness or no knowledge

Source: Moohan, 1993, p26

that environmental issues under the classification of Social Studies are considered to be medium order skills alongside IT, European Issues, Valuations, Economics and Finance, Legal Studies and Professional Practice. As Table 4 illustrates, Social Studies are perceived to be skills or areas of expertise requiring medium order skills across the different divisions. Based on the evidence it is safe to assume that environmental issues are considered of the same importance as some of the other important areas of surveying.

At this juncture it is therefore pertinent to discuss how environmental issues should be included in the overall curricula of surveying courses.

Table 3 Skills or areas of expertise requiring a higher order ranking in General Practice Surveying

Skills or areas of expertise	Ranking
Interpersonal skills	High
Current social policy issues	Medium
Price determination of property interests	High
Investment appraisal	High
Statutory valuations	High
Property Law	High
Property/Estate Management	High
The system of town and country planning in the UK	High
Development Appraisal	High

Source: Moohan, 1993, p69

Table 4 Selected skills or areas of expertise requiring medium order skills (understanding of principles and professional practice prior to APC)

Skills or area of expertise	Division						
Social Studies	BS	QS	GP	PD	RP	MS	LS
The wider and longer term implications of land use activities	+	+	+	+	+	+	+
The main social factors and influences that affect the built environment	+	+	+	+	+	+	-
Trends in social structure	+	-	+	+	-	+	-
Current social policy issues	-	-	+	+	+	-	-
Economic and Finance							
Applied Taxation	+	+	+	+	+	+	+
Investment Appraisal	+	+	+	+	+	+	-
Information Technology							
The use of wordprocessors and computer spreadsheet	+	+	+	+	+	+	+

Source: Moohan, 1993, p40,42,43

ENVIRONMENTAL EDUCATION OF SURVEYORS

As outlined above, a growing number of surveying educators and practitioners have recently argued that their profession will soon go into deeper recession or become marginal to the new world if it is not advanced beyond its traditional sphere of activities. In particular, they are concerned that the prevailing approaches to surveying education are sending out surveyors are are inadequately prepared for the emerging economic conditions. Whilst this anxiety is about the appropriateness of the education and training of new graduates, it is further complicated by the ever increasing encroachment of the surveyors' professional domain by other professionals. The new environmental agenda poses a more difficult challenge. However, dictums such as fostering interdisciplinary education and specialisation for better practice are more easily vocalised than put into operation. The RICS has recognised this and as demonstrated above, there appears to be a growing awareness of green issues by members of the Institution. This section is therefore devoted to a debate on how environmental issues could be incorporated in the education of surveyors.

A series of in-depth semi-structured interviews were held with a number of educators and representatives of institutional investors as well as professional firms in order to evaluate their perception of environmental issues as they relate to surveying education and training. The interviews were focused around a number of general and specific issues ranging from the role of surveyors in tackling environmental issues, the areas of professional practice where surveyors could be expected to apply their knowledge on environmental issues, the degree of specialism required, the stage at which environmental issues should be included in surveyors' education and the role of the RICS in encouraging the environmental education of chartered surveyors. The respondents from the education bodies were mainly heads of surveying schools or directors of undergraduate and post-graduate surveying courses. Although the majority of the interviews were based around General Practice and Planning and Development, a number of Building Surveying and Quantity Surveying interviews were also included.

Discussions with educators and surveying firms suggest that it is widely recognised that surveyors have a role to play in tackling environmental issues as part of their overall activity. A number of respondents note that for surveyors to take a lead would be a natural role for responsible professionals due to the nature of their activity, i.e. the management of landed property. Surveyors are seen to be in a position to bring together the range of environmental issues related to property in their capacity as project managers, with the help of specialists. Their understanding of the different aspects of landed property enables them to consider a wide range of potential implications in terms of investment risk, financial costs and liabilities. A senior academic observes that 'surveyors have historically been concerned with environmental issues as estate surveyors, land agents and stewards of greater estates and therefore have a knowledge of the negative impact of drainage on forestry but this role has changed with the move from this premise due to concern with the market and investment pressure'. The same is argued by Thompson (1965) in linking the concepts of stewardship and custodian to the responsibility of surveyors.

The changing nature of legislation and development is placing additional demands on surveyors particularly on the extent to which they can provide advice. Paice (1995) argues that rural surveyors may be familiar with agricultural pollution risks and relevant legislation. However, environmental assessment is already starting to have a greater bearing on rural land management, not only where planning applications or diversifications on agricultural land are involved, but in terms of far wider interpretations of the word 'pollution'. According to a leading rural surveyor specialising in planning and environmental consultancy, David Hoare (cited in Paice, 1995) 'Land agents may know about the potential pollution on dairy farms and so on but doubts that all of the profession are aware of all relevant legislation. He suggests as many as 80% do not know it in detail'.

What could be deduced from the above account is that there is a majority who believe that surveyors should have knowledge of and involvement with environmental issues. They warn against surveyors offering specialist environmental advice; specialist advice should be given by environmental consultants. Evidence from other research on attitudes of investment institutions suggests that most of these institutions utilise desktop environmental surveys to provide an initial warning of potential problems (Lizieri et al. 1995). Many of the surveyors interviewed believe that this type of service could be provided by surveyors. The results of the interviews also suggest that surveyors should have a general knowledge and skill in related areas of professional practice. Larger professional firms and the institutional investors have begun to incorporate environmental issues in their business activities. Only a few large surveying firms provide any distinctive environmental services but they recognise this as a growing market. A leading firm of surveyors claim that environment related services account for 7% of their annual turnover. However, they believe the value added in terms of creating opportunities for other work has been much greater. Amongst the institutions interviewed, one holds the view that:

> '*Surveyors should be able to read technical reports and understand asset implications and if need be call in expert advice. Furthermore surveyors should appreciate the environmental performance of buildings, performance of environmentally friendly buildings, likelihood of future legislation and their implications*'

The same investor believes that because of investment risk, surveyors are as close to environmental issues as other professionals and therefore they decided to put together an environmental policy for the whole group. They assert that 'environment is part of life - everything you touch is the environment'.

COURSE DESIGN AND CURRICULUM GREENING

The plethora of research work on greening of higher education has helped to identify a number of approaches how environmental education could be incorporated in higher education.

Table 5 Areas of professional practice to which surveyors are expected to apply knowledge on environmental issues.

Areas of professional practice	Highly relevant	Relevant
Contaminated land	+	*
Valuation	*	+
Environmental liabilities	*	+
Investment risk	+	*
Tenant risk	*	+
Environmental management	+	+
Environmental statement	+	*
Environmental legislation	*	+
Environmental audits of buildings	+	*
Environmental audits of land and property	+	*
Waste management & disposal	*	+

+   The majority view
*   The minority view

Shirley Ali Khan (1995) proposes that as students spend a considerable length of time in an educational institution, therefore they are affected by the environmental practices of that institution. She argues that 'the promotion of sustainable practice across higher education curricula must therefore be set in the context of a wider strategy, akin to the corporate environmental strategies which an increasing number of companies are adopting' (Ali Khan, 1995, p3). In this context a university owns a great deal of landed property, has a multi-million pound turnover and as part of its overall activities it utilises energy, water and numerous other materials. The Toyne Report also emphasises this role and recommends that 'every FHE institution should formally adopt and publicise, by the beginning of the academic year 1994/5, a comprehensive environmental policy statement together with an action plan for its implementation' (Toyne, 1993). But apart from organisational practice, the dilemma facing course directors and lecturers is one of how to determine the exact method of introduction and delivery of environmental education. In order to determine what course of action should be adopted it is important that educationalists concern themselves with the fact that 'students are the Professionals of the future'. Wood (1994) argues that 'it is to be expected that today's students will take with them into industry a well-developed sense of environmental responsibility and that those already in the workplace respond to conscience as well as market forces and legislation in wanting to adopt greener practices'. Furthermore they should be in a position to link their own individual responsibility and the implications of their actions as consumers of resources and producers of waste. The message is therefore how can individuals relate 'global concepts' to 'individual lifestyles'.

Students on built environment courses in general and the surveying discipline in particular are closer to issues of global warming, conservation, materials production, consumption and management than many other students. Therefore what they learn in their curricula will directly have an impact on their actions as decision makers. There are three methods of introduction and delivery of environmental education in student curricula. Shirley Ali Khan (1995) defines these as "addition", "incorporation" and 'engagement'.

The first approach deals with environmental issues marginally, based on the assumption that the environment has little relevance to the main course discipline. The method of delivery in this scenario is by way of giving a special guest lecture or a special option or a specialist course, thus treating the environment as a separate issue.

The second approach makes the assumption that the environment has `some relevance' to the discipline (Ali Khan, 1995) thus attempting to contribute to the environmental knowledge base. In this scenario, environmental connections are made to the existing discipline of the course by an evolutionary process. By adopting this approach some revision of the course is undertaken with either the introduction of a new slant on established theory and practice or by rejection of such theories and practices.

The engagement approach, however, significantly changes the shape of the course. Whilst the incorporation approach does not have an environmental action agenda, the engagement approach is associated with an action agenda for the environment. In this scenario the environmental education is considered part and parcel of the overall curricula and treated similarly to other issues such as the use of information technology, increased European participation through improved language skills and the development of enterprise skills (Ali Khan, 1995).

The engagement approach could itself be tackled in different ways. Firstly through 'limited engagement' on the premise that disciplines have something to offer to the environment. The ultimate aim is problem solving even when the result turns out to be partial or no solution at all. The 'full engagement' approach, however, makes the assumption that disciplines have much to offer the environment with the goal of a sustainable future thus moving from the limited problem-centred approach to a vision-directed approach.

Wood (1994) outlines five methods of integration and delivery. Modular approaches occur through the introduction of optional or compulsory modules. Whilst optional modules are offered to students on a number of different courses, often on subjects related to their main course, compulsory modules are planned to include subject relevant material in the course. Syllabus alteration involves modification to all course elements to provide an environmental perspective. The success of this approach is conditional on staff developing new course materials and learning new knowledge. The student project approach on the other hand allows students in-depth investigation of the link between environmental and other aspects of their chosen topic. Thus this approach has a general awareness raising capacity.

Experiential learning may incorporate a number of options including, site visits to exhibitions, attendance at a public inquiry or visits to specialist centres such as the Centre for Alternative Technology. The final method of integration is 'environmental learning outcomes' when students are expected to achieve an expected level of environmental competence at the end of their course. Otter (1993) observes that 'in many employers' eyes, the development of broader graduate competence is more important than subject knowledge including the development of environmental competence'. Wood (1994) advocates, rather than the more traditional, process related bench-marks, using learning outcomes which take more account of prior learning and work-based experiences. Table 6 outlines the main approaches to environmental education.

Examination of a typical surveying degree in estate management leading to APC in the General Practice Division provides a suitable platform for debate on where and at what level environmental issues could be incorporated in the overall curricula. Table 7 provides an example of such a curriculum. The units of study in the example reflect the requirements of the RICS Reports and Guidelines by the inclusion of core subjects which provide both Non Surveying Specific and Surveying Specific skills. The unit or module based structure of the course provides the university with the flexibility to deliver the required subjects. For universities it is a more efficient use of resources and for students it offers a degree of choice through selection of options on a variety of subjects including foreign language, environmental law and business studies. In terms of environmental education, although the course provides the student with the option of an Environmental Law Unit, it does not offer a recognisable theme of green issues throughout the overall curriculum. As the previous section suggests environmental issues are applicable to any of the units of study outlined in Table 7.

To decide where and at what level to deliver environmental elements is problematic. The educators are faced with the difficulty of deciding how much new material should be brought to the area, whether the postgraduate courses should provide this and finally what would be the reaction of the Surveying Courses Board.

Of the educational institutions interviewed, the majority do not have dedicated environmental units as part of their current curricula, but some claim to have incorporated elements of green issues into the relevant units or modules of study.The units in Table 8 are cited as most relevant to environmental issues.

Table 6 Approaches to curriculum greening

Approach	Characteristics	Delivery method
Additive	Assumption that the environment has little relevance to the existing concerns of the host discipline	Single seminars/lectures
Incorporation	Assumption that the environment has some relevance to the discipline only providing a new slant on established theory or practice	Project
		Dedicated modules
Engagement	May significantly change the shape of disciplinary enquiry. Possible problems (too much change in emphasis)	Change to curricula
Limited engagement/problem solving	Assumption that disciplines have something to offer the environment. The starting point is the environmental issues to which disciplinary contribution to solutions are sought	Cross-curricula greening
Full engagement	Assumption that disciplines have much to offer the environment. The starting point is the positive goal of a sustainable future, a goal to which disciplinary contributions are actively sought	New specialist courses

Table 7 The contents of a four year sandwich BSc (Hons) Estate Management course

Level 1 Units	Level 2 Units	Sandwich Year	Level 3 Units	Teaching & Assessment Method
Investment and Valuations	Property Management & Development	Professional placement to provide opportunity for application of theory within a supervised working environment	European Studies including visit	Directed learning through formal lectures, seminars, workshops, group project work and desk and field research
Valuations	Valuation and Law		Development and Planning	
Law	Property Law		Valuation and Law	Assessment by an aggregate of exam and course work
Economics	Land Economics	The year counts towards one year of professional practice required for RICS membership	Management & marketing	
Land Policy and Planning	Estate Management		Property investment analysis	
Construction Technology 1	Construction Technology 2			
Information Management	Integrative Project Studies		Dissertation	
Project Studies	Options from French, German, Spanish, Conservation Studies, Property Research, Information Analysis		Options from Finance & Accounts, Environmental Law, Corporate Strategy & Business Law, French, German & Spanish	
Options from French, German and Spanish				

Source: South Bank University, 1995. Undergraduate Prospectus 1995/96, p150.

The number of schools having integrated units bringing together different streams of environmental issues related to property is very limited at present, although many have expressed the view that this may offer the right approach. The stage at which environmental issues could be introduced in the curriculum either during university education or prior to APC is a major topic of discussion. Currently the RICS requires surveying graduates or non cognate entrants to gain two years of professional practice prior to Assessment of Professional Competence. This is an important stage in a surveyor's education and training and may offer a potential opportunity to ensure whether candidates have acquired an awareness of environmental issues before becoming associates of the RICS. However, only a minority of those interviewed support this idea. The introduction of dedicated environmental units or modules at undergraduate level receives most support, much more than specialist post graduate-courses.

Table 8 Units of studies incorporating relevant environmental issues

Name of units	Ranking in terms of importance (1= most important, 10= least important)
Construction Technology	1
Conservation Studies	2
Land Policy & Planning	3
Development & Planning	4
Property Law	5
Environmental Law	6
Land Economics	7
Estate Management	8
Investment & valuation	9
Property investment & development	10

Table 9 The stage at which environmental issues should be included in surveyors' education

Stage of introduction	Ranking in terms of importance
At the university whilst students on surveying courses by adoption of dedicated units or modules	1
At the post-graduate level by adoption of specialist courses	2
After RICS membership through CD programme	3
Prior to RICS membership and during APC	4

There are also resource issues in terms of available specialist staff to teach the new subjects and deciding on how to acquire the required knowledge, whether through recruitment of new staff or staff development. Much of this depends on how well institutions have adopted the recommendations of the Toyne Report. The Toyne Report

(1993) recommended that all institutions in HFE should develop policies in relation to three areas of 'environmental responsibility'.

1. Curriculum: the report envisages that all students should have some basic education in environmental issues rather than only those on specialist 'environment' courses.
2. CPD and Consultancy: institutions should develop a policy to provide updating training and consultancy to industry and commerce in this area.
3. Institutional impact: institutions should adopt environmentally responsible policies in respect of their own operations - in order to 'practise what we teach'. This would include a wide ranging set of policies in the area of waste management, recycling, energy saving, using 'environmentally sound' materials, etc.

Many universities have adopted elements of the above recommendations but the degree of success has varied at different institutions. Of the surveying schools interviewed, only a few have well developed environmental action plans and course curricula reflecting environmental issues. As outlined in an earlier section, modular teaching allows cross-curricula teaching with environmental options often being delivered by members of staff from other schools in the university. But the success of this depends on co-operation and internal communication between different schools. In other cases, some schools have seconded members to private environmental consultancies to attain the relevant knowledge. Upon their return the staff have been able to pass on their knowledge to other members of staff by short course presentations and transfer of their knowledge in teaching and course design. The senior academic in charge of this particular surveying school holds the view that 'such developments need a long lead-time as well as resources. The school has developed specialism through staff development policy and build up the expertise in environmental issues'. This has been a particularly successful strategy in providing opportunities for diversification in the face of falling numbers of recruits. It has allowed the development of new specialist post-graduate courses in Environmental Management for Business and Commerce.

THE RICS - ROLE AND RESPONSIBILITIES

As a professional institution with a long history, the RICS's influence over the education and training of its members is unique. The RICS is therefore responsible for seeking out new services for its members and promoting surveying as an enlightened and caring profession involved with the protection of the environment. This should result in enhancing the image of surveyors. A senior academic sees the RICS 'as a professional institution developing a role in an evolving market for diversity from the current situation in the surveying profession'. So far the RICS has not published any formal policy on environmental education, but it does have skill panels dealing with contaminated land and environmental management. Amongst other professional institutions directly involved with the built environment, the Royal Town Planning Institute (RTPI) in 1985 published a statement urging the consideration of environmental issues in planning education, but does not have an up to date policy. The Royal Institute of British Architects (RIBA) published an 'Initial Statement on Global Warming and Ozone Depletion' in 1990 offering guidelines on energy efficiency and design. The Institute of Housing does not have any environmental education policy either. As discussed in earlier parts of this chapter, although the RICS does not have any policy at present, it has launched a number of initiatives with published outcomes offering some guidelines on contaminated land, waste management and pollution. The majority of those interviewed whether the educators or firms believe that the RICS should play a more proactive role in encouraging environmental education for chartered surveyors. With the gradual encroachment on surveyors' traditional areas of work by other professions and in the light of the current condition of the property market, many believe environmental issues could create new opportunities for surveyors. Others believe that the RICS's opposition to the proposed Register of Contaminated Land was negative and therefore a more proactive and positive role is needed. A leading investor comments that 'when it comes down to it, the valuer is at the core of Chartered Surveying business and evaluation of risk is at the core of his skills' and urges the RICS to help its members put this subject into context by:

1. Making sure valuation is dealt with in a professional and competent manner
2. Contributing to the debate, and
3. Influencing the education and training process by being more vigorous through accreditation to make courses take on board the relevant environmental issues.

Another investor notes the over-capacity of the market for surveying services and recommends that unless they have the necessary qualifications and expertise they stand to lose out to competition. Other educators hold the view that as far as the profession is concerned they are looking for new business and client base: 'we could therefore provide a better training for such surveyors to offer sound services and advice for the ever increasing environmentally conscious client and the RICS could help increase this client base'. Some educators observe that as the RICS no longer examines, its role in the education process is changing. At the moment there appears to be some conflict between the prescriptive and the permissive approaches. They prefer the RICS to take a lead in the promotion of environmental issues in practice: as the academics are already converted. Others expect the RICS to identify the areas of work where environmental issues impact on General Practice Surveying and urge the Surveying Courses Board to decide on what should be covered in courses. The degree to which the RICS should be prescriptive in its policy towards accreditation of courses is subject to disagreement amongst educators. As pointed out some prefer the RICS to be more proactive whilst others want it to take a supportive role. The majority however agree that the RICS should be:

1. Proactive and take a lead role in influencing the education of surveyors regarding environmental issues,
2. Proactive in promoting CPD courses, and

3.   Stimulating/ funding research in relevant areas.

Whilst many educators prefer the RICS to have a less prescriptive role in general and view centre accreditation as an advantage rather than individual course accreditation, at the same time they 'do not want to endanger their own accreditation'. They assert that the RICS has no choice but to keep its membership informed and be at the forefront of environmental issues. A number of professional and investment firms point out that they already have difficulties recruiting graduates with sufficient knowledge of environmental issues. If this is the case at present, when there is limited demand for environmental services, the situation could become far worse in the near future. Therefore the role of the RICS in promoting environmental education for surveyors is paramount.

Evidence from the surge in property related environmental research suggests that surveyors have become increasingly involved with environmental issues. Therefore the RICS is required to act in its advisory role to promote a systematic feed-back mechanism in education and training of chartered surveyors. The recent initiatives by the RICS including the Land Quality Statement Note:

'remind chartered surveyor of the skills that are needed in advising owners, developers, lenders or investors in recognising the seriousness of environmental liability' (Wilbourn, 1995).

Wilbourn, the author of Land Contamination Guidance for Chartered Surveyors, states that 'the RICS is not suggesting that the chartered surveyor should become an environmental consultant. The "added value" of the chartered surveyor relies on the development of skills to interpret advice provided by other experts in applying this to a property activity'. This is a view supported by the majority of those interviewed during the course of this research.

CONCLUSION

This chapter set out to establish the current position of environmental education for the property profession. It sought to evaluate the views of different interest groups involved with the education and training of chartered surveyors. Current property market conditions have created an over-capacity for surveying services and many believe environmental issues could provide diversity for the surveying profession and new opportunities for surveyors. The results of interviews support previous research in highlighting the impact of environmental issues on the property market and the growing need for environmental services. In terms of course design and the incorporation of green issues in surveying education, it has been demonstrated that at present only a few surveying schools have introduced dedicated and recognisable environmental units/modules to their overall curricula. The majority of other schools claim to have included environmental elements in the relevant core modules.

However, there is a need to provide cross-curricular greening in surveying education. It is recognised that the RICS should play a more proactive advisory, but less prescriptive, role in encouraging greening of property education and training. The nature of surveying as a broad based profession poses particular difficulties in recommending what should be incorporated in the overall surveying curricula without endangering the quality of the core educational requirement.

However, all signs point to a need for surveyors to have adequate knowledge of environmental issues to be able to have the skills to offer advice on potential asset risk to owners, developers, lenders or investors. As the 'Education Policy - A Strategy for Action' (RICS, 1994b) recognises, there is no single model of educational provision that can hope to satisfy all of these constituents'. In the greening of property education, the ultimate task should be to consider that today's students are tomorrow's policy and decision makers, managers and professionals. Our aim is to provide the necessary education which would help them recognise the positive goal of sustainable development relating global concepts to individual lifestyles. The inclusion of environmental education should not be seen as a separate element but as an essential part of the overall curricula. Whilst the RICS should play an important role in its professional mission, educators share this responsibility in terms of course design and delivery.

REFERENCES

Ali Khan, S. (1993). *Environmental Issues in Further and Higher* Education: *Coomb Lodge Report.* Further Education and Development Agency, London.

Ali Khan, S. (1995). *The Environmental Agenda, Taking Responsibility - Overview* London, Pluto Press.

Betts, D. (1994). Greening the campus. *NATFHE Journal.* Autumn.

Davies, C. (1993). Contaminated land - professional liability. Paper presented at the *RICS Cutting Edge Conference.*

Estates Times. (1995). *RICS takes on polluted land.* 1266, February 10.

Fraser, W., Crosby, N., MacGregor, B. & Venmore-Rowland, P. (1994). Education of GP Surveyors - confusion worse confounded. *Estates Gazette,* 9403, January 15.

Fraser, W., Crosby, N. & MacGregor, B. (1995). Professional education - Issues of quantity and quality. *Estates Gazette,* 9507, February 18.

Hall, B. (1995). Europe's environmental law set to cause bigger headaches. *Property Week,* 9 February, p27.

Lay Committee (1991). *Market requirements of the Profession.* A report to the Royal Institution of Chartered Surveyors, November.

Lizieri, C., Palmer, S., Charlton, M. & Finlay, L. (1996). *Valuation Methodology and Environmental Legislation.* RICS Research Papers, London.

Moohan, J.A.J. (1993). *Procedures and guidelines for course accreditation supplementary advisory notes: Postgraduate programmes.* RICS, London.

NATFHE (1994). Working for a greener world. *NATFHE Journal.* Autumn.

Otter, S. (1993). *Learning Outcomes: Issues in Course Delivery and Design,* a conference for the staff of De Montfort University, Leicester.

Paice, C. (1995). Environmental Assessment - challenging territory for rural practice. *Estates Gazette.* July 1, 9526, pp126.

Parsa, A. (1993). *Impact of environmental issues on the UK property industry.* Paper presented at the 3rd Australasian Real Educators' Conference - *Future Directions in Property* 26-29 January 1994, Sydney, Australia.

Parsa, A. (1995). The greening of property education. *Chartered Surveyor Monthly,* July/August, pp27.

Pott, S. (1995). *People in Property - The inaugural address of the President of the RICS* given on 11 July 1995 at the RICS Headquarters in London.

RIBA (1990). *Initial Policy Statement on Global Warming and Ozone Depletion.* RIBA, London.

RICS (1995a). *Land Contamination Guidance for Chartered Surveyors.* RICS, London.

RICS (1995b). *Environmental Management and the Chartered Surveyor.* RICS, London.

RICS (1994a). *Chartered Surveyors: The Property profession, Degrees and Diplomas Accredited by the RICS.* RICS, London.

RICS (1994b). *Education Policy: Trends in Recruitment, Education, Training and Career Development - A Strategy for Action.* July, RICS, London.

RTPI (1985). Environmental Education - Policy Statement, 19 June, *The Planner.* August.

Smith, P.F. (1995). *Notes on the question of a sustainable future - an introductory information document.* University of Sheffield EC SAVE Programme, Sheffield.

Thompson, F. (1965). *Chartered Surveyors - the growth of the profession.*Routledge, London.

Toyne, P. (1993). *Environmental Responsibility in Further and Higher Education.* Department For Education, London.

Toyne, P. (1994). Green shoots on Campus - a year of progress?. *Greening Universities,* 1(3). October Newsletter of the Greening of Higher Education Council, Oxford.

Turner, N. (1994). *Environmental Management Systems and their use as a risk reduction strategy in property investment.* Paper presented at *the RICS Cutting Edge Conference* 2-3 September.

Wilbourn, P. (1995). Chartered surveyors and land quality statements. *Estates Gazette,* 9509, March 4, p311.

Wood, S. (1994). *The Environmental Agenda, Taking Responsibility - Built Environment.* Pluto Press, London.

10 The Risk from Radon in Abandoned Mine Workings

G. R. BARKER-READ & R. A. FARNFIELD
The University of Leeds, Leeds, UK

INTRODUCTION

Exposure to radioactive substances, both in the natural environment and during the course of employment activities, is of great concern because of the potential that ionising radiation has for causing damage to biological materials. For many years the effects of high doses of radiation have been well known and understood, as have the benefits to be gained from the controlled exposure to radiation in the medical context. What is less well understood is the effect of the dose received through continual or episodic exposure to low levels of radiation - both naturally occurring and man-made - accumulated by the population at large as they go about their daily business. Such doses and exposures may well be negligible; but on occasions they are not and there may exist an increased risk to health associated with some work or leisure activities. Caving, pot-holing, entry into disused mines and other underground chambers is one area where there exists the potential to accumulate a greater than average dose of radiation due to exposure to naturally-occurring radon gas and its decay progeny.

The work reported in this paper was conducted as a means of demonstrating how the magnitude of the risk posed by one type of activity, namely the entry into a disused metalliferous mine, may be quantified and how it varies throughout the year. This was achieved by installing radon monitoring equipment in an abandoned lead mine in North Yorkshire, together with facilities for sampling water issuing from the mine and the climatic conditions inside and outside of the mine. The paper describes the nature of radon and the radon 'daughters', their geographical and geological distribution in the UK, the current legislation relating to exposure at work, the techniques and instrumentation available for measurement, and concludes with a presentation and discussion of the results obtained during the mine survey.

WHAT IS RADON?

At the end of the last century Owens and Rutherford (Rutherford, 1900a,b),

The Environmental Impact of Land and Property Management, Edited by Yvonne Rydin
© 1996, The Royal Institution of Chartered Surveyors

while investigating the characteristics of thorium salts, discovered that a radioactive substance diffused continuously from the salt. A similar phenomenon was observed from actinium. In the following year Dorn (Dorn, 1901) identified a gaseous emission from salts of radium. Elster and Geitel (Elster and Geitel, 1901) demonstrated the presence of radioactivity in the atmosphere which was later shown to be due to radon and thoron daughters. In 1902 Rutherford and Soddy were able to condense radon and Soddy, five years later, proved that radon was a member of the noble gas family (Ramsey, 1907). Thus within a decade of Becquerel's discovery of radioactivity, the general nature of atmospheric radioactivity was established and concern about the effects of continuously inhaling radioactive air was beginning to be expressed (Rutherford, 1907).

Radon is a noble gas with 29 known isotopes. All are radioactive and three of them (^{219}Rn, ^{220}Rn and ^{222}Rn) occur in nature as members of the primordial actinium, thorium and uranium series, respectively. The name 'radon' is customarily reserved for ^{222}Rn; ^{219}Rn and ^{220}Rn are usually called actinon and thoron respectively, after their series parents. All three isotopes are alpha emitters; their half-lives are given in Table 1 (after Duggan et al., 1990).

Table 1: Half-lives of radon isotopes

Series	Isotope	Common name	Half-life
Actinium	^{219}Rn	Actinon	3.96 seconds
Thorium	^{220}Rn	Thoron	55.6 seconds
Uranium	^{222}Rn	Radon	3.82 days

The parent atoms of these series can be found in all natural materials and so all three radon isotopes are released into the air from the surfaces of rocks, soils and building materials. Principally because of their short half-lives, the atmospheric concentrations of actinon and thoron are much less than that of radon, thus the latter is the isotope of principal interest here.

In general, geology has a controlling influence on the magnitude of the radon emission problem since it is from the rock that the gas escapes. The source of the radon is ^{238}U which has a half-life of approximately 4.5 billion years, thus radon has been released into British rocks since the radioactive minerals were incorporated into them from Precambrian times onwards. The radon exhaled from the earth's surface into the free atmosphere is rapidly dispersed and diluted by vertical convection and turbulence. There is less opportunity for dispersion in confined air spaces such as houses (particularly their basements and sub-floor voids), underground service ducts, mines and caves. In such areas the radon concentration increases with decreasing ventilation rate. Since it is highly soluble in water radon may also enter confined spaces in solution, being released to the atmosphere upon evaporation or aeration.

The risk to health arising from inhalation of radon gas was first investigated following the recognition of an increased incidence of lung cancer in several groups of miners in uranium mines, metalliferous mines and even in mines where there were apparently no radioactive minerals. Central European pitchblende miners' deaths probably attributable to the inhalation of radon and its daughter products date back as far as the sixteenth century (Agricola, 1556), although it was not until the 1920s that the link was suggested (Ludwig and Lorenser, 1924); it was finally proved in 1940 (Evans and Goodman, 1940). More recently (Bale, 1951) it is the radon daughters (^{218}Po, ^{214}Pb, ^{214}Bi and ^{214}Po) in the atmosphere which have been shown to be markedly more hazardous to health than radon gas itself and are now regarded as the primary cause of the observed lung cancers. These are all electrically charged, chemically active solids which rapidly attach to particulate matter in the atmosphere, mainly the respirable dust fraction. The radon daughters may then be inhaled and become trapped in the lungs and undergo several decay steps, including alpha decay, before being exhaled. It is the short-lived radon daughters which, when inhaled, deliver the alpha radiation dose to bronchial tissue that is implicated in radiogenic lung cancer. An epidemiological study of uranium miners in the USA (Archer et al., 1973) established a quantitative relationship between the number of deaths resulting from lung cancer and the cumulative exposure to alpha particle energy due to radon decay.

Since radon gas is soluble in water it can also enter the body by ingestion. It dissolves in the blood with a partition coefficient of 50%, thus up to 20 litres of radon can exist in the body (Fremlin, 1989). It is soluble in fat and so can remain available for some time. Should an atom of radon decay within the blood or body tissue the decay products, being solids, cannot easily be excreted and further radioactive decay can cause considerable cellular damage.

The hazard is not confined to uranium mines. In 1964 (de Villiers & Windish, 1964) airborne radioactivity was shown to be the main cause of a high incidence of lung cancer amongst Canadian fluorspar miners and in 1970 a similar link was highlighted for British haematite miners (Boyd et al., 1970). Strong and co-workers published the results of a national survey of the level of radon and its daughters in British non-coal mines (Strong et al., 1975). They were able to detect radon present in the atmosphere in every one of the 52 mines sampled, in some cases at alarmingly high levels. A clear correlation was established between the level of radon in the mine atmosphere and the efficacy of the ventilation. Radon and its daughters is usually detectable also in coal mines, although at much lower concentration due principally to the necessarily high ventilation rate needed to adequately dilute firedamp (Gardner, 1995; British Coal Corp., 1990).

The recognition that radon and its daughters are also responsible for a major part of the radiation dose received by the population at large is fairly recent. Since the 1950s evidence of enhanced radon levels in dwellings and other buildings has been gathered resulting, in 1977, in the publication of the International Commission on Radiological Protection document known as Publication 26 (ICRP, 1977) which recognised, for the first time, that there may be unhealthy accumulations of radon in domestic properties which might have to be controlled. Subsequently, it became evident that radon in the home was a major hazard in some parts of the UK, in particular south-west England, parts of north Wales, Derbyshire, North Yorkshire and the Highlands, Grampian and Southern Uplands of Scotland. A full discussion of the survey data, quantification of the risk and suggested remedial action can be found in Duggan (Duggan et al., 1990) whilst O'Riordan (O'Riordan, 1990) details the advice of the National Radiological Protection Board.

Units of measurement

Radon gas concentration is measured in terms of activity per unit volume. The unit of activity, the Becquerel, is defined as being one disintegration per second, thus radon concentration is expressed in Bq/m^3. The activity of a source is proportional to the number of radioactive atoms present and so diminishes with time in an enclosed atmosphere as the decay occurs. The half-life of a source is the time taken for the activity, or the original number of atoms present, to fall by 50%.

Radon daughter concentration is measured in terms of the alpha energy released per unit volume (J/m^3) or in a special unit - the Working Level (WL) - where 1WL is that concentration of short-lived daughters of ^{222}Rn in equilibrium with $3700Bq/m^3$ of radon gas. This is equivalent to that concentration of decay products of radon which releases a total of $2.08 \times 10^5 J/m^3$ $(1.3 \times 10^5 MeV/l)$ of alpha energy. The Working Level is an exposure level and not a dose rate and 1WL was, historically, thought to be the safe limit of exposure.

Cumulative exposure to radon decay products is commonly expressed in units of Working Level Months (WLM). An exposure of 1WLM can be taken to be received by a person working in a radon daughter concentration of 1WL for 170 hours (a 'standard' month's working time). Expressed another way:

Exposure in WLM   =   (WL x h)/170   where h is the time of
                                                exposure in hours

The maximum permissible annual exposure to radon daughters is not internationally agreed upon. However, exposure to 4.8WLM of radon daughters gives rise to a committed effective dose equivalent of 50mSv, which is the current annual (whole body) radiation dose limit for employees aged 18 or over (IRR, 1985). When there is exposure to radon daughters there will normally be additional exposure to other naturally occurring radiation, which must also be included in the dose assessment. To simplify monitoring procedures a practical exposure limit of 4WLM is used and radon daughters only are measured, rather than the cumulative effect of all natural radiation sources. This, then, implies that a steady concentration of 0.33WL is considered safe to work in for 170 hours each month.

Regulatory framework

In 1986 the Ionising Radiations Regulations came into force and laid down requirements for the 'protection of persons against ionising radiation arising from any work activity'. The Regulations were made under the Health and Safety at Work etc. Act 1974 and are administered by the Health and Safety Executive. They apply to radon when there is an exposure of employees to an atmosphere containing the short-lived daughters of ^{222}Rn at a concentration in air, averaged over any 8 hour period, of greater than 6.24 x 10^{-7}J/m^3, which is equivalent to 0.03WL.

At higher exposures the Regulations set out the actions that are required to be taken:

< 0.03WL defined as 'Not work with ionising radiation'; no action required.

0.03 - 0.10WL 'Supervised area' declared; Health and Safety Executive to be informed and some regularity of monitoring to be carried out.

0.10 - 0.30WL 'Controlled area' declared; workers and environment to be continuously monitored. Access to controlled areas must be restricted to classified workers or to persons working to a written system of work which can be shown to limit their exposure appropriately.

0.33WL 'dose limit'; the maximum permitted level of continuous exposure giving rise to an annual whole body dose of 50mSv. (Since the Regulations were introduced the ICRP has recommended that this limit be reduced to 20mSv averaged over a defined period of five years, with the added provision that the effective dose should not exceed 50mSv in any one year (ICRP, 1991). In due course the Regulations are likely to be amended to incorporate these lower values.)

Work places where such exposures may occur include: any working mine; tourist mines and caverns; other underground workplaces including civil engineering works; and ground level workplaces and basements where radon can accumulate after permeating to the surface from underlying rock containing uranium.

Surveyors are commonly required to carry out their duties in such places.

No places above ground floor level and no large industrial workplaces at ground level have been found with radon daughter concentrations above 0.03WL. A HSC Approved Code of Practice document 'Exposure to Radon' (Health & Safety Commission, 1988) gives advice on what to do if the Regulations apply and also gives guidelines on the regularity of monitoring required at various radon daughter concentrations.

To put the problem in perspective, the 1975 NRPB survey of British non-coal mines (Strong et al., 1975) found radon daughter concentrations above the

0.03WL threshold present in underground workings of every mineral except one (a salt mine), and in excess of 1WL in several tin and lead mines. The highest concentration measured was 19WL in a Cornish tin mine. For all other minerals (haematite, ironstone, fluorspar, gypsum, limestone, calcspar, honestone, fireclay, ballclay, slate and barytes) the usual concentration lay between these values. A clear correlation was established between radon daughter concentration and ventilation rate. The radon daughter concentration present in the atmosphere within coal mines rarely exceeds the 0.03WL threshold (except in South Yorkshire) due to the large volumes of air circulated through the workings for firedamp control.

Adequate ventilation or a limitation on the time spent by persons in an unsafe concentration are the only methods by which exposure to radon daughters can be controlled since it is impracticable to prevent the emission of radon from the strata. In working mines the problem can (and by law must) be controlled so that no person is exposed to unsafe levels of radon at their workplace. In abandoned and disused mines, natural caverns, underground chambers and service ducts any ventilation is almost always natural and usually weak, thus much higher concentrations of radon can accumulate (Lively and Krafthefer, 1995).

Natural ventilation effects arise due to differences in the density of air within and outside of a mine. Such a difference may be brought about by changes in air temperature. It is a fact that the temperature of the rock, and hence of the air, deep in a mine is constant throughout the year whilst that on the surface of the earth varies considerably with the seasons. Consequently, the amount and direction of natural ventilation depend on the time of year and, to a lesser extent, the time of day. It would be reasonable to assume, therefore, that the strength and direction of natural ventilation flow in a mine would have a controlling influence on the airborne radon concentration present in various parts of that mine and, hence, the risk to the health of persons entering therein.

In a similar manner the amount of radon dissolved in mine drainage water could reasonably be expected to be affected by the rate at which the water enters the mine (thus, indirectly, the amount and distribution of rainfall) and the rate at which it flows out. It can thus be seen that, if we assume that the supply of radon from the strata is effectively constant and limitless, its concentration in the mine air and water is effectively determined by the prevailing weather and the time of year.

MEASUREMENT PROTOCOLS

From the foregoing discussion it is apparent that there are, in fact, two areas of concern associated with emanation of radon - exposure to the radon gas itself and exposure to the particulate daughter products. Both are alpha emitters and it is this characteristic which is commonly used in their determination. To make the situation more difficult, it is required to measure either the activity (equivalent to the concentration) of radon gas in Bq/m^3 or to measure the exposure, in WL, to the decay products. Fortunately the two are related by a simple expression.

With an effectively constant source, radon gas will reach equilibrium with its decay products in a period of three to four hours, such that the radon and all its daughters will achieve the same activity. In this state the atmosphere is said to be fully 'aged'. A disequilibrium factor (f) represents the degree to which this equilibrium has been reached and can be defined as:

$$f \quad = \quad 3700 \text{ x [Radon daughter concentration (WL)] / [Radon gas concentration } (Bq/m^3)]$$

Theoretically, f can lie between zero and one. In well-ventilated working mines the value of f is usually found to be in the range 0.2 to 0.7; values between 0.4 and 0.8 can be expected in naturally-ventilated excavations and between 0.5 and 1.0 for non-ventilated chambers. Once the disequilibrium factor for a particular site has been established by concurrent measurement then only the radon gas or its progeny concentration need be measured. Clearly, the value of f is substantially affected by the ventilation conditions and can thus fluctuate with time.

A number of techniques have been developed for determining radon gas or radon daughter concentration. The radon gas may be separated from its particulate daughters by simple filtration - the gas will pass through the filter whilst the solids are retained on its surface. This can be achieved either by actively pumping the atmosphere through a filter (glass-fibre or membrane) or by allowing it to passively diffuse through. Obviously, the latter is a much slower process and is not suitable for rapidly changing concentrations, but it has the advantage of requiring no power source.

Once separated, a known volume of radon gas is allowed to decay and the alpha activity is measured either after a few minutes, when it is due solely to the first decay product (^{218}Po), or when equilibrium has been established. A number of different types of instrument for making integrated or continuous measurements have been developed, but the passive detector is becoming favoured. A radon detection system of this type known as the E-PERM system, developed by Rad Elec Inc of Frederick, Maryland, USA, has been deployed in Gilfield mine for nearly two years. This system is based upon the use of a digital electrostatic voltmeter to measure the difference in potential on the surface of an 'electret' in response to ionisation due to alpha radiation.

The electret is a small wafer of Teflon that holds a stable electrostatic charge. This is placed in an electrically conductive plastic ion chamber of known volume into which the radon is allowed to diffuse via a filter. When a radon atom radioactively disintegrates, the nucleus emits an alpha particle which ionises the air in the chamber. A little cloud of free electrons and positively charged (ionised) oxygen and nitrogen atoms is left in the path of the alpha particle. The charged atoms and electrons are now moved by the electrostatic field established by the electret. Negative ions are attracted to the positively charged electret; when they reach the electret surface they reduce its electrostatic potential by a measurable amount. By determining the voltage loss over a known time interval it is possible to determine the average concentration of radon in the ion chamber, which must be equal to that in the atmosphere outside of the chamber.

One of the advantages of the E-PERM system is its versatility. Using suitable accessories it may also be used to determine radon dissolved in water, by releasing radon from a measured volume of water into a closed volume of relatively radon-free air; samples of soil can similarly be assessed for radon flux over a given area or ^{226}Ra content. Personal dosimetry is possible using small ion chambers worn on the clothing; special chambers are also available for measurement of thoron, tritium and beta radiation.

Radon daughter concentration is measured by filtering a known volume of the atmosphere and then counting the alpha activity on the filter at a measured time after sampling. The Kuznetz method and the Tsivoglou method are well-established procedures for determining Working Level; both require a 5 to 10 minute sampling period followed by activity measurement lasting up to 40 minutes.

More recently, electronic instruments, sometimes known as 'instant' Working Level meters, have become available. Typical of such instruments is the Radon Working Level Meter manufactured by Thomson & Nielsen Electronics of Ottawa, Ontario, Canada, used in this study. It is a hand held, direct reading, portable unit with its own rechargeable internal 6V battery and pump; alternatively it can be powered externally for prolonged periods. The pump draws air at a rate of one litre per minute through a 0.8 micron membrane filter, upon which the radon daughter particles are deposited. The radon gas is vented back to atmosphere. The alpha activity on the filter is observed electronically and the instrument displays the total alpha count digitally which, in a pre-determined time, is related to the Working Level.

A programmable data logging unit is available for long-term regular sampling and results storage; downloading facilities to a computer are also available. By suitably programming the unit the variation in Working Level may be established over time. For example, in this study the mean Working Level was established on an hourly basis.

FIELDWORK UNDERTAKEN AT GILFIELD MINE

Gilfield Mine is situated to the north of the B6265 Pateley Bridge to Grassington Road at Greenhow Hill in the county of North Yorkshire. The Greenhow Hill area was one of the country's leading lead mining fields of the 18th and 19th centuries and its history, together with that of Gilfield Mine itself, part of the Cockhill and Sunny Side Lead Mines group, is well documented (Dickinson, 1985; Dunham and Stubblefield, 1945; Raistrick, 1973; McFarlane, 1984).

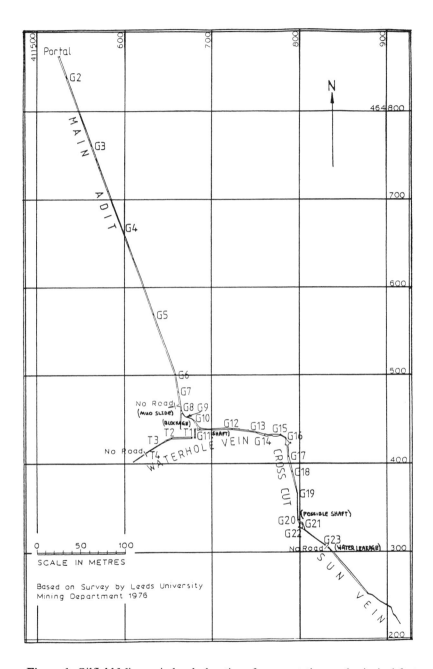

Figure 1: Gilfield Mine main level - location of survey stations and principal features

Gilfield Mine is one of several adits extending towards the south under Greenhow Hill. Between 1782 and 1789 the adit was driven straight in a south-easterly direction into the hillside for a distance of some 425m to intersect the Waterhole Vein system, Figure 1. The tunnel was about 2m in height and 1.5m in width and was driven on a rising gradient of about 1 in 150 to assist drainage. Upon meeting the Waterhole Vein system open overhand stull stopes were worked to the east and west both above and below adit level. At this point the adit is about 100m below ground level; surface elevation being about 400m AOD.

The main drive continued to the east in the orebody for a distance of some 120m until a cross-cut was driven to intersect Sun Vein about 100m to the south. From here the workings continue to the south in the Sun Vein system for a considerable distance and connect with other mines on several horizons. The original mine workings were very extensive and penetrated a variety of rock types including Upper Carboniferous gritstones, shales with inferior coals, and Lower Carboniferous limestone. The radon is believed to be associated with fine-grained deep water shales and bituminous limestones, although its source has not been positively established.

Significant underground mining of lead ore continued at Gilfield until about 1880 when imported ores from Spain and Australia proved to be more economically attractive. Between 1920 and 1935 the mine was worked sporadically for lead and fluorspar; the last mining to take place was the robbing of fluorspar-rich pillars and pockets in the early 1940s. Several attempts were made in the 1970s to re-work the extensive waste heaps for fluorspar; limited recent drilling of the deeper parts of Sun Vein has similarly proved fruitless.

Since 1974 the mine and about 5 acres of surrounding land has been leased by The University of Leeds and a field centre constructed for undergraduate teaching and research. In particular, it has proved extremely valuable in the teaching of practical surveying (surface and underground), geological mapping, geochemical prospecting and instruction in environmental monitoring and measurement techniques. Access to the mine is strictly controlled since it is registered with the Mines and Quarries Inspectorate and is thus subject to the provisions of the Mines and Quarries Act, 1954. It is also designated as a Controlled Area as defined in the Ionising Radiations Regulations, 1985.

The workings now accessible consist of a main free-draining adit of some 700m length (from the portal to point G23 in Figure 1) and an internal ladderway at G11 leading up into a stulled stope some 80m in length. Apart from the main adit no surface connection is accessible; the numerous old shafts having collapsed or been filled for safety reasons, and the ventilation and drainage routes through the old workings are unknown. There is good evidence to suggest that the workings below adit level are permanently flooded. Radon investigations have been performed in the mine since 1976 (Young, 1977; Senaratne, 1978; McFarlane, 1984).

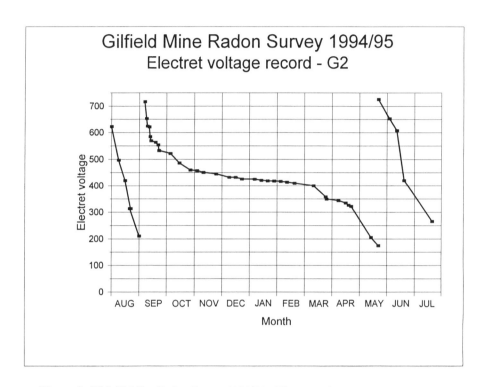

Figure 2 Gilfield Mine Radon Survey 1994/95 - Electret voltage record-G2

Gilfield Mine is typical of the many thousands of abandoned metalliferous mines, caves and natural caverns which exist in the UK. It is excavated in hard rock types thus is generally stable (apart from the artificial stope floors which are supported on old wood stulls, most of which are now rotten) and was, prior to the erection of a substantial gate at the mouth of the adit, frequently visited by cavers and other members of the public ignorant of the presence of ionising radiation. Whilst it is unlikely that such persons received dangerous doses of radiation in the mine, their annual dose would undoubtedly have been raised above the 2.2mSv average dose received from the 'background' radiation level in the UK The major contributor to natural exposure in the UK is radon (ambient outdoor concentration is about $3Bq/m^3$) and its decay products which typically account for about 55-60% of the average background radiation, although both the background level and the contribution attributable to radon vary geographically.

The fact that the majority of such subterranean voids exist in areas of the country also known to be rich in radon, such as Cornwall and Devon, the Peak District, North Yorkshire etc., is no coincidence but a function of the geology. For example, the richly mineralised granites of the south-west, for centuries mined for tin and copper, are also rich in the decay-series nuclides and have a fractured structure which facilitates the migration of the gaseous radon. Similarly, the limestones of the Pennine orefields, exploited for ores of lead, zinc, copper and non-metallic minerals such as fluorite, gypsum and barytes, are host to extensive natural fractures, cave systems and ancient mine workings.

Since the mine has been used for practical underground surveying for many years a number of permanent survey stations have been established, Figure 1, at which parameters such as tunnel cross-sectional area, rock temperature and ventilation rate are known with reasonable accuracy. E-PERM electret ion chambers were installed at stations G2, G11 and G23 in the main adit and at T2 in the stope. The voltage of each was read at approximately weekly intervals over the period August 1994 to July 1995 from which the weekly average airborne radon concentration was derived. Figure 2 shows a typical record of electret voltage and the corresponding radon concentration is plotted in Figure 3.

The Thomson & Nielsen Working Level Meter and a data logger were installed at station G11 and set to record hourly average values of radon daughter Working Level. The data logger was downloaded at approximately weekly intervals when the external batteries were replaced. By determining both the radon gas and radon daughter concentration at G11 simultaneously the disequilibrium factor (f) could be established and its variation with ventilation conditions determined.

At station G2 a Prosser AVM2 hot-wire anemometer was installed to measure air velocity at hourly intervals. The output from this sensor was cabled out of the mine to a second data logger. The product of the air velocity and the cross-sectional area of the measuring station, assuming reasonably stable conditions, yields the air volumetric flow. Platinum resistance thermometers were installed at G2 and G4 to record air dry-bulb temperatures, again connected to the data logger outside of the mine.

A portable weather station was installed at the mouth of the adit which was set up to determine air dry-bulb temperature, barometric pressure and rainfall at hourly intervals; this was also connected to the data logger. With a knowledge of the air temperatures and barometric pressure it is possible to determine the air density inside and outside of the mine and thus to deduce the natural ventilation pressure, which is the driving force responsible for the air flow.

Drainage water flow from the mine was measured using a specially constructed V-notch weir in the stream leading from the adit. Although a satisfactory means of continuously measuring water flow over the weir was developed, the unusual lack of rainfall during the measurement period made accurate calibration of the system impossible.

Passive monitoring for radon gas

Figures 3, 4, 5 and 6 show the airborne radon concentration at the four measurement stations. In each case there is a distinct period during which the radon concentration is consistently low - this is the winter months of October to March inclusive. This is most obvious at station G2 and is caused by the natural ventilation effect. During this period the average daily air temperature within the mine (constant at approximately 7°C at G4 but known to be 9°C deeper in the mine) is higher than that of the outside, air thus the latter is more dense and exerts a greater pressure. A rough calculation shows the difference in barometric pressure along the accessible length of the main drive to be of the order of 2 to 3Pa at most, yet the net result is a steady flow of air into the mine during the winter months in response to the pressure difference.

The air entering the mine contains a relatively low level of radon (the ambient background concentration for this area) hence the measured level at G2 is, as expected, negligible during winter. As one follows the drive into the hillside the concentration of radon increases marginally through G11 to G23 due to the pickup of radon emitted from the adit walls and that released from solution as the drainage water is naturally aerated as it flows out of the mine. Even so, the 'winter' level of radon at G23 rarely exceeds 500Bq/m^3. The small peak in concentration discernible in March, most noticeable at G2, is the result of a temporary reversal of air flow direction, such as is described later.

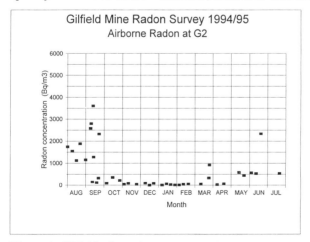

Figure 3 Gilfield mine radon survey 1994/95: airborne radon at G2

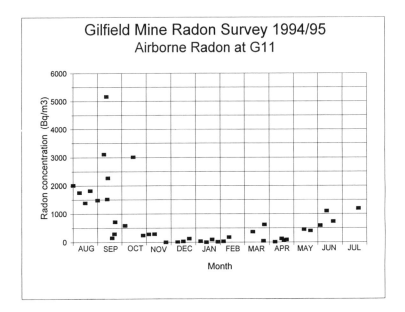

Figure 4 Gilfield mine radon survey 1994/95: airborne radon at G11

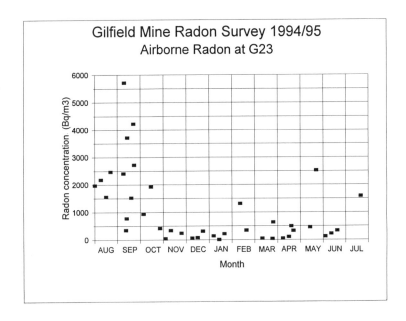

Figure 5 Gilfield mine radon survey 1994/95: airborne radon at G23

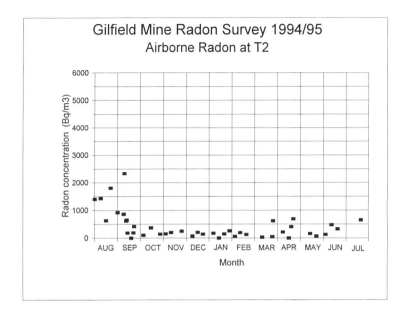

Figure 6 Gilfield mine radon survey 1994/95: airborne radon at T2

From April/May until early October of each year the situation is reversed, i.e, the average daily air temperature outside of the mine exceeds that within, resulting in lower air density outside and a constant flow of air along the main drive and out of the adit mouth. Since the air present in the main level of the mine has come from within the deeper workings, it has ample time to accumulate a rich concentration of radon, at times over ten times the typical winter level. Even the air entering the main drive from the upper level stopes, via T2, is richer in radon than in the winter months. The two airstreams join in the vicinity of G11 and flow out of the mine as a single flow; the diluting effect is discernible in the generally lower concentrations recorded at G2.

Active monitoring for radon daughters

Figure 7 shows the outcome of the continuous radon daughter monitoring exercise performed at station G11. The quality of the data gathered, although generally acceptable, was not as good as had been expected due to problems with the data logger. However, a general outline and discussion of the results are presented here.

The monitoring system was installed on 13/04/95 and continuous recording commenced automatically at 00:00 on 14/04/95. The equipment performed satisfactorily from its installation until 13:00 on 11/05/95 when a fault developed in the data logger which resulted in total loss of Working Level data until the equipment was repaired and re-installed on 25/05/95. Throughout this period the other measured parameters (air temperature, pressure, etc.) were recorded without loss of data on the second logger outside of the mine. For a subsequent period of some 106 hours the logger recorded an incomplete record of radon daughter concentration; this is reflected in the straight line section through late May. In fact, what had happened was that during the period that the logger was defective the radon daughter concentration in the air had risen to such a level that the span of the logger analogue input had been exceeded. The effect was to overload the logger, resulting in 'clipping' of the data peaks (the troughs were not affected). This, of course, was not apparent until the logger was downloaded on 31/05/95, whereupon its dynamic range was reset to provide a suitable span. From this point until monitoring ceased at 12:00 on 05/07/95 (some 1978 hours from first installation) an accurate continuous record was obtained.

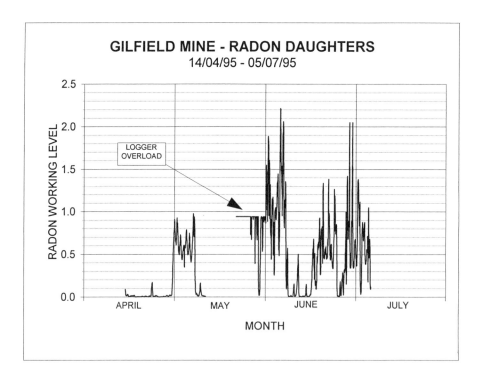

Figure 7 Gilfield mine - radon daughters 14/04/95 to 05/07/95

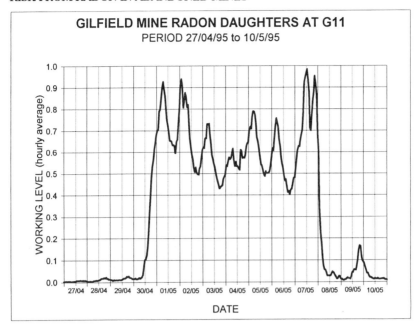

Figure 8 Gilfield mine radon daughters at G11 period 27/04/95 to 10/05/95

Observation of the graph reveals that the radon daughter level in the mine air was effectively negligible until the end of April 1995. From early on 30/04/95 to about midnight on 07/05/95 radon daughter concentration varied between 0.4WL and 1.0WL, falling again to a very low value thereafter. The initial rise and final decline in level during this time are extremely abrupt and will be discussed in more detail later. Periods of low activity are also to be seen between 07/06/95 and 15/06/95, and again between 24/06/95 and 27/06/95. During the periods of high activity very rapid fluctuations in Working Level can be identified; the peak value of 2.28WL was attained at 08:00 on 05/06/95. It is interesting also to note that on each of the five occasions when the radon daughter concentration exceeded 1.8WL, the peak value was reached between 07:00 and 08:00 hours.

Figure 8 shows the radon concentration for the two week period from 00:00 on 27/04/95 until 24:00 on 10/05/95. Until 09:00 on 30/05/95 the radon daughter concentration in the airstream does not exceed a value of 0.03WL (the minimum threshold stipulated in the Ionising Radiations Regulations 1985 for the declaration of a 'Supervised Area'), yet in the following 23 hours it rises to 0.93WL, which is nearly three times the acceptable exposure limit. Similarly, in the 13 hour period from 21:00 on 07/05/95 to 10:00 on 08/05/95 the level drops rapidly from 0.95WL to 0.03WL. Assuming that the rate of radon emission from the strata is constant, and that there is no mechanical interference with the ventilation, the only possible cause for such rapid changes must be a reversal of air flow direction throughout the entire mine due to natural causes.

Inspection of Figure 9 shows the reason for the switch in flow direction. In this figure the thick solid line indicates the temperature of the air outside of the mine, the thin solid line the temperature of the air at station G2 and the circles show air temperature at G4, which is almost constant at 7.0°C. This is, in fact, the strata temperature at this point, indicating that rock and air are in thermal equilibrium at this distance from the mouth of the mine.

Looking first at the outside air temperature, the diurnal variation is quite large and unusually high daytime temperatures are reached for this early time of year. Until mid morning on 30/04/95 the air is flowing into the mine and the record at G2 shows an air temperature which follows closely, though at reduced magnitude, the external temperature variation. Air entering the mine is always cooler than that at G4. From mid morning on the 30th the G2 air temperature ceases to fluctuate and slowly approaches that at G4 indicating that a flow reversal has taken place over a surprisingly short time span. Figure 10 shows the magnitude of the ventilation current at G2; the change in direction and the surprisingly short time this takes to achieve are clearly visible.

Ventilation conditions within the mine remain reasonably stable until early afternoon on 07/05/95 when air temperature at G2 is seen to climb above that at G4, indicating that a second reversal of flow direction has occurred and the G2 record once again begins to reflect the variation in external air temperature. Air is again flowing into the adit at this point and the radon concentration measured at G11 is reduced to very low levels. Similar strong causal relationships can be demonstrated for other time periods.

From the concurrent measurements of radon gas and radon daughter concentration at G11 the disequilibrium factor for this site (f) was determined to be approximately 0.8 when the flow was stable in 'summer' conditions, i.e. out of the adit. This value was, in fact, found to be highly variable with time indicating that the atmosphere at G11 is not fully 'aged' and that no reliance can be placed on the use of the disequilibrium factor as a means of determining radon daughter concentration from radon gas activity, or vice versa, at this site.

Figure 11 indicates the results of the radon in water survey; a degree of caution is necessary in the interpretation of these results since the flows involved are very small due to the lack of rainfall during the measurement period. However, the level of dissolved radon in the water at G23 would appear to be generally higher than that measured at G2, as could reasonably be expected. The fall in activity of the water as it flows along the adit and out of the mine can be attributed to the process of aeration which occurs - the radon desorbs from the water and enters the air. There appears to be no discernible variation in dissolved radon content with time of year. Further measurement will, no doubt, shed more light on the air/water relationship.

It is interesting to consider briefly what happens to the radon-rich air when it leaves the mine portal. Radon gas is some 7.7 times as dense as normal atmospheric air. If climatic conditions are such that immediately outside of the

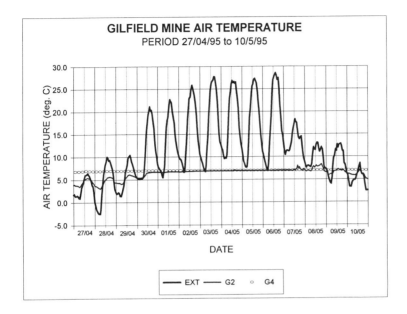

Figure 9 Gilfield mine air temperature period 27/04/95 to 10/05/95

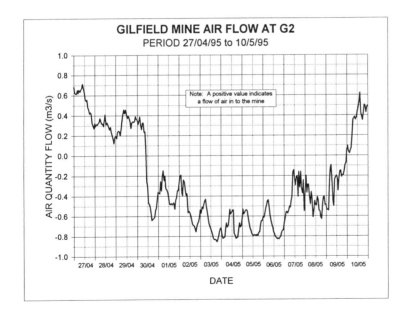

Figure 10 Gilfield mine air flow at G2 period to 27/04/95 to 10/05/95

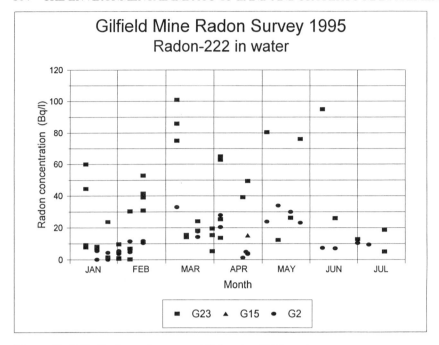

Figure 11 Gilfield mine radon survey 1995: radon-222 in water

mine there is little or no air turbulence (necessary to dilute and disperse the radon), then it has a tendency to 'pool' in hollows in the ground and may even flow downhill following the water courses. Thus the risk of exposure to enhanced levels of radon is not necessarily confined to the mine workings themselves, but may also extend to some distance around the mine opening. Further measurement is necessary to define the limits of such affected areas.

THE RISK TO HEALTH

So what is the risk to the health of persons entering the mine? Clearly, during the winter months when the flow of air is into the mine along the main level towards the deeper workings, the concentration of radon and its daughters in the air is negligible and any activities in the currently accessible part of the mine can be classed as 'not working with ionising radiations' (less than 0.03WL). However, in the summer months, when the air flow direction is reversed, radon-rich air flows out of the mine along the main level and the airborne concentration rises considerably above the 0.33WL 'dose limit'. The highest level recorded in this study was 2.2WL. If this concentration were sustained continuously then a visitor would only be allowed to remain in the mine for about 31 hours before requiring medical supervision under the terms of the Regulations. There is clearly a good case to be made for insisting that all persons entering a Supervised or Controlled

Area should wear a personal dosemeter to indicate their exposure to alpha activity; if not, then area monitoring must be performed to define the level of risk and a tight control kept on time spent underground to prevent persons accumulating high doses of alpha radiation.

The problem of unintentional exposure to radon is not confined to the old lead mines of North Yorkshire, of course. It is present, to a greater or lesser degree depending on the geology and ventilation efficacy, in every underground chamber, void or excavation. Potholes, show caves and tourist mines, for example. Unless they are mechanically ventilated, the same criteria will apply in these caverns as in Gilfield Mine - the risk will be greatest when the outside air temperature exceeds that within the void, resulting in a flow of radon-rich air out from the deeper recesses of the cavern. This will occur in the summer months which, unfortunately, is precisely the time of year when the majority of tourists wish to enter them. Similarly, there would appear to exist a risk that persons living in the vicinity of old mine workings or natural caverns may well be exposed to higher levels of radiation than the general public, either through inhalation of air flowing from the mine or through consuming or using radon-rich water.

ACKNOWLEDGEMENTS

The authors, and the Department of Mining and Mineral Engineering, University of Leeds, acknowledge with gratitude the financial support received from The Royal Institution of Chartered Surveyors for the pursuit of this work. Reference to particular measurement instruments should not be construed as endorsement of those systems.

REFERENCES

Agricola, G. (1556). *De Re Metallica*. Translated by H.O. and L.H. Hoover. Dover Publications Inc, New York.

Archer, V.E., Wagoner, J.K. & Lundin, F.E. (1973). Lung cancer among uranium miners in the United States. *Health Physics*, **25**, 351-371.

Bale, W.F. (1951). Memorandum to the files, March 14 1951: hazards associated with radon and thoron. *Health Physics*, **38**, 1061 (1980).

Boyd, J.T., Doll, R., Faulds, J.S. & Leiper, J. (1970), Cancer of the lung in iron ore (haematite) miners, *British Journal of Industrial Medicine.*, **27**, 97-105.

British Coal Corporation (1990). *Radon Decay Products Monitoring in Coal Mines, Report on ECSC Research Project 7260-03/052/01*. CEC, Brussels.

de Villiers, A.J. & Windish, J.P. (1964). Lung cancer in a fluorspar mining community. 1: Radiation, dust, and mortality experience. *British Journal of Industrial. Medicine*, **21**, 94-109.

Dickinson, J.M. (1985) *Mines and T'Miners: A History of Lead Mining in Airedale, Wharfedale and Nidderdale.* 3rd imp. Office Liaison Ltd, Sheffield.

Dorn, F.E. (1901). Von radioaktiven substanzen ausgesandte Emanaton, *Abhandlung uber die Naturorschende Gesellschaft.* Halle. **23**, 1.

Duggan, M.J., Al-affan, I.M., Ballance, P.E., Henshaw, D.L., Lillicrap, S. & Thomas, D.C. (1990). *The Risks from Radon in Homes - A Report of a Working Group of the Institute of Radiation Protection,* HHSC Handbook No 5, H & H Scientific Consultants Ltd Leeds.

Dunham, K.C. & Stubblefield, C.J. (1945). Stratigraphy, structure and mineralisation of the Greenhow mining area, *Quarterly Journal of the Geological Society of London,* **100**, 209-268.

Elster, von J. and Geitel, H. (1901). Uber eine fernere Analogic in dem elektrischen Verhalten der naturlichen un der durk Becquerelstrahlen abnorm gemachten Luft. *Physikalische Zeitschrift* **2**, 590-593.

Evans, R.D. & Goodman, C. (1940). Determination of the thoron content of air and its bearing on lung cancer hazards in industry. *Journal of. Industrial. Hygene. and Toxicology.*

Fremlin, J.H.(1989). *How radon affects us.* Paper presented at an open meeting organised by the Institute of Radiation Protection, 11 January.

Gardner, A.E. (1995). Radon emissions from abandoned mines. *The Safety and Health Practitioner,* November.

Health and Safety Commission (1988). Exposure to Radon - The Ionising Radiations Regulations 1985, *Approved Code of Practice, Part 3.* HMSO, London.

ICRP (1977). International Commission on Radiological Protection, Recommendations of the International Commission on Radiological Protection, ICRP Publication 26, *Annals of the ICRP,* **1**: 3.

ICRP (1991). International Commission on Radiological Protection, Recommendations of the International Commission on Radiological Protection, ICRP Publication 60, *Annals of the ICRP,* **21**: 1-3.

Ionising Radiations Regulations, 1985 (1985). *SI No 1333,* HMSO, London.

Lively, R.S. & Krafthefer, B.C. (1995). ^{222}Rn variations in Mystery Cave, Minnesota, *Health Physics,* **68**: 590-594.

Ludwig, P. & Lorenser, E. (1924). Untersuchungen der Grubenluft in den Schneeberger Gruben auf den Gehalt an Radiumemanation, *Zeitschrift fur Physik,* **22**: 178-185.

McFarlane, J. (1984). Radon investigations at Gilfield Mine Teaching and Research Centre, Greenhow. *Transactions of the. Leeds Geological. Association.,* **10**(6): 68-88.

O'Riordan, M.C. (1990). Human exposure to radon in homes; recommendations for the practical application of the Board's statement. *Documents of the NRPB 1,* **1**: 17-32.

Raistrick, A. (1973). *Lead Mining in the Mid-Pennines*, D. Bradford Barton Ltd.,Truro.

Ramsay, W. (1907). Note, *Nature.* **76**, 269.

Rutherford, E. (1900a). Note. *Philosophical. Magazine.* **49**: 1.

Rutherford, E. (1900b). Note. *Philosophical. Magazine.* **49**: 161

Rutherford, E. (1907). Some cosmical aspects of radioactivity. *J. R. Astronomical. Society. Canada,* May-June, pp 145-165.

Senaratne, C.P.J. (1978). Ventilation and Radon Survey at Gilfield Mine, Greenhow, N. Yorkshire, Postgraduate Diploma dissertation, University of Leeds, (unpublished).

Strong, J.C., Laidlaw, A.J. & O'Riordan, M.C. (1975). Radon and its Daughters in Various British Mines, *National Radiological Protection Board, Report 39*, HMSO, London

Young, P.A. (1977). Exposure to radon daughters at Gilfield, Private communication.

INDEX

Abandoned mine-workings, risk from
    radon in, 165-87
Agenda 21, 35
Air, radon-rich, movement of, 183-4
All-risks-yield (ARY figure),
    102,108,109,119
Alterations/conversions, 86-7,95
    see also Refurbishment
Amenity valuation, 62,63
Approved Document L, 81,84,93,95
    1995 revisions, 83,85
    changes to requirements affecting
        existing buildings, 85-6
    and energy conservation,
        88,89,*90,91,92*

Bixby Ranch case, California, 109
Blueprint Target Groups, 42
Boilers, high efficiency, 92
Brundtland Report, 2,14,15
BS5750 (Quality Management Systems),
    17,23,32
BS7750 (Environmental Management),
    23,25,28-9,32,55,62
Building Act (1984), 93
Building Control Authorities, and
    standards of work, 93-4
Building Employers Federation (BEF),
    42,44
Building Regulations, 81,84-5,96
    and change to office use, 86-7
    Danish, 93
Building Regulations (1995), 69,94-5
    SAP target incorporated, 69
Building Regulations (Amendment
    Regulations) (1994), 81
Building surveyors, and Approved
    Document L, 93

Buildings
    'as built', poor design and/or
        workmanship, 94
    change of use, 85-6
Business, local, and Agenda 21, 44,47-8
Business and Industry Targeting Groups,
    43

Calculation of worth, 97,110-11,117,*119*
Carbon dioxide emissions, 81,*83,91*,93,94
    from domestic, public and commercial
        buildings, 82
    reduction of, 65-6,68-9,71,77
Catchment area management plans, 138
Catchment areas, 132,136,137
Chartered surveyors, 13,24
    education needs of, 146-51
    and the land development process, 143-
        4
    see also Quantity surveyors; Rural
        practice surveyors; Surveyors;
        Valuers
City Challenge, 37
Client advice, 25,33
Clients, 6-7,13,60
    commissioning environmental work, 58
    needs of, and environmental concerns,
        57-8
Climate change, 52-3
'Climate Change Programme'(UK),
    81,82,84
Coast protection schemes, 126
Coastal developments, division of
    responsibilities, 128
Coastal Forum, 137
Coastal groups, UK, 136-7
Coastal management plans, 138

Index compiled by Mrs Connie Tyler BA.